AF577133

Polyamine Protocols

METHODS IN MOLECULAR BIOLOGY™

John M. Walker, SERIES EDITOR

90. **Drug–DNA Interaction Protocols,** edited by *Keith R. Fox, 1998*
89. **Retinoid Protocols,** edited by *Christopher P. F. Redfern, 1998*
88. **Protein Targeting Protocols,** edited by *Roger A. Clegg, 1998*
87. **Combinatorial Peptide Library Protocols,** edited by *Shmuel Cabilly, 1998*
86. **RNA Isolation and Characterization Protocols,** edited by *Ralph Rapley, 1998*
85. **Differential Display Methods and Protocols,** edited by *Peng Liang and Arthur B. Pardee, 1997*
84. **Transmembrane Signaling Protocols,** edited by *Dafna Bar-Sagi, 1998*
83. **Receptor Signal Transduction Protocols,** edited by *R. A. J. Challiss, 1997*
82. **Arabadopsis Protocols,** edited by *José M Martinez-Zapater and Julio Salinas, 1998*
81. **Plant Virology Protocols,** edited by *Gary D. Foster, 1998*
80. **Immunochemical Protocols (2nd. ed.),** edited by *John Pound, 1998*
79. **Polyamine Protocols,** edited by *David M. L. Morgan,* 1998
78. **Antibacterial Peptide Protocols,** edited by *William M. Shafer, 1997*
77. **Protein Synthesis:** *Methods and Protocols,* edited by *Robin Martin, 1998*
76. **Glycoanalysis Protocols,** edited by *Elizabeth F. Hounsel, 1998*
75. **Basic Cell Culture Protocols,** edited by *Jeffrey W. Pollard and John M. Walker, 1997*
74. **Ribozyme Protocols,** edited by *Philip C. Turner, 1997*
73. **Neuropeptide Protocols,** edited by *G. Brent Irvine and Carvell H. Williams, 1997*
72. **Neurotransmitter Methods,** edited by *Richard C. Rayne, 1997*
71. **PRINS and *In Situ* PCR Protocols,** edited by *John R. Gosden, 1997*
70. **Sequence Data Analysis Guidebook,** edited by *Simon R. Swindell, 1997*
69. **cDNA Library Protocols,** edited by *Ian G. Cowell and Caroline A. Austin, 1997*
68. **Gene Isolation and Mapping Protocols,** edited by *Jacqueline Boultwood, 1997*
67. **PCR Cloning Protocols:** *From Molecular Cloning to Genetic Engineering,* edited by *Bruce A. White, 1996*
66. **Epitope Mapping Protocols,** edited by *Glenn E. Morris, 1996*
65. **PCR Sequencing Protocols,** edited by *Ralph Rapley, 1996*
64. **Protein Sequencing Protocols,** edited by *Bryan J. Smith, 1996*
63. **Recombinant Proteins:** *Detection and Isolation Protocols,* edited by *Rocky S. Tuan, 1996*
62. **Recombinant Gene Expression Protocols,** edited by *Rocky S. Tuan, 1996*
61. **Protein and Peptide Analysis by Mass Spectrometry,** edited by *John R. Chapman, 1996*
60. **Protein NMR Protocols,** edited by *David G. Reid, 1996*
59. **Protein Purification Protocols,** edited by *Shawn Doonan, 1996*
58. **Basic DNA and RNA Protocols,** edited by *Adrian J. Harwood, 1996*
57. **In Vitro Mutagenesis Protocols,** edited by *Michael K. Trower, 1996*
56. **Crystallographic Methods and Protocols,** edited by *Christopher Jones, Barbara Mulloy, and Mark Sanderson, 1996*
55. **Plant Cell Electroporation and Electrofusion Protocols,** edited by *Jac A. Nickoloff, 1995*
54. **YAC Protocols,** edited by *David Markie, 1995*
53. **Yeast Protocols:** *Methods in Cell and Molecular Biology,* edited by *Ivor H. Evans, 1996*
52. **Capillary Electrophoresis:** *Principles, Instrumentation, and Applications,* edited by *Kevin D. Altria, 1996*
51. **Antibody Engineering Protocols,** edited by *Sudhir Paul, 1995*
50. **Species Diagnostics Protocols:** *PCR and Other Nucleic Acid Methods,* edited by *Justin P. Clapp, 1996*
49. **Plant Gene Transfer and Expression Protocols,** edited by *Heddwyn Jones, 1995*
48. **Animal Cell Electroporation and Electrofusion Protocols,** edited by *Jac A. Nickoloff, 1995*
47. **Electroporation Protocols for Microorganisms,** edited by *Jac A. Nickoloff, 1995*
46. **Diagnostic Bacteriology Protocols,** edited by *Jenny Howard and David M. Whitcombe, 1995*
45. **Monoclonal Antibody Protocols,** edited by *William C. Davis, 1995*
44. ***Agrobacterium* Protocols,** edited by *Kevan M. A. Gartland and Michael R. Davey, 1995*
43. **In Vitro Toxicity Testing Protocols,** edited by *Sheila O'Hare and Chris K. Atterwill, 1995*
42. **ELISA:** *Theory and Practice,* by *John R. Crowther, 1995*
41. **Signal Transduction Protocols,** edited by *David A. Kendall and Stephen J. Hill, 1995*
40. **Protein Stability and Folding:** *Theory and Practice,* edited by *Bret A. Shirley, 1995*
39. **Baculovirus Expression Protocols,** edited by *Christopher D. Richardson, 1995*
38. **Cryopreservation and Freeze-Drying Protocols,** edited by *John G. Day and Mark R. McLellan, 1995*
37. **In Vitro Transcription and Translation Protocols,** edited by *Martin J. Tymms, 1995*
36. **Peptide Analysis Protocols,** edited by *Ben M. Dunn and Michael W. Pennington, 1994*
35. **Peptide Synthesis Protocols,** edited by *Michael W. Pennington and Ben M. Dunn, 1994*
34. **Immunocytochemical Methods and Protocols,** edited by *Lorette C. Javois, 1994*

METHODS IN MOLECULAR BIOLOGY™

Polyamine Protocols

Edited by

David M. L. Morgan

King's College, London, UK

Humana Press Totowa, New Jersey

999 Riverview Drive, Suite 208
Totowa, New Jersey 07512

This publication is printed on acid-free paper. ∞
ANSI Z39.48-1984 (American Standards Institute)
Permanence of Paper for Printed Library Materials.

Cover illustration: Fig. 1 in Chapter 19, "Measurement of Polyamine Efflux from Cells in Culture," by Heather M. Wallace and A. Jill Mackarel

Cover design by Patricia F. Cleary.

For additional copies, pricing for bulk purchases, and/or information about other Humana titles, contact Humana at the above address or at any of the following numbers: Tel.: 973-256-1699; Fax: 973-256-8341; E-mail: humana@mindspring.com; or visit our Website: http://humanapress.com

Printed in the United States of America. 10 9 8 7 6 5 4 3 2 1

Library of Congress Cataloging in Publication Data

Main entry under title:

Methods in molecular biology™.

Polyamine protocols/edited by David M. L. Morgan.
p, cm.—(Methods in molecular biology; 79)
Includes index.
ISBN 0-89603-448-8 (alk. paper)
1. Polyamines—Laboratory manuals. 2. Morgan, M. L. II. Series: Methods in molecular biology (Clifton, NJ); 79.
[DNLM: 1. Polyamines—analysis. 2. Polyamines—metabolism. W1 ME9616J v. 79 1998/QU 61 P778 1998]
QP801.P638P626 1998
572'.548—dc21
DNLM/DLC
for Library of Congress 97-37499
CIP

Preface

Putrescine and spermidine are ubiquitous in living organisms. Spermine, third of the three most commonly occurring natural polyamines, is probably present in all eukaryotes but is rare (or nonexistent) in prokaryotes. Polyamine residues are constituents of many compounds found in plants and insects. Putrescine, spermidine, or spermine-containing alkaloids are found in many plants, nonproteinaceous spider and wasp toxins contain polyamine residues, and glutathionyl–spermidine conjugates have been found in some pathogenic microorganisms.

In most cells polyamines are the products of a highly regulated biosynthetic pathway. It is not clear whether the elaborate regulation of polyamine synthesis is a consequence of their essential role(s) in cellular differentiation and development, or part of a defense mechanism to prevent overaccumulation of compounds that are toxic in excess. In addition to their biosynthetic capability, many cells also possess transport systems for polyamines that respond to intracellular polyamine levels, and other stimuli, and are regulated by mechanisms that are at present incompletely defined.

Two routes of polyamine catabolism have been identified in mammalian cells, a biodegradative route and a recycling pathway. The relative importance of these pathways and their overall regulation is only partially resolved. What is clear is the widespread occurrence of a variety of polyamine-oxidizing enzymes in animals, plants, bacteria, and fungi. Polyamine catabolism, by whichever route, results in the formation of aminoaldehydes as intermediates. These compounds, containing one or more amino groups that will be positively charged at physiological pH, and an aldehyde function, are highly reactive. Indeed, aminoaldehydes have been shown to be cytotoxic to a wide variety of cell types. It is not yet clear whether these compounds have any biological function (although a regulatory role has been postulated) or are merely unstable and rapidly degraded intermediates in the polyamine catabolic pathway.

Intracellular polyamine concentrations vary throughout the cell cycle. An increase in polyamine synthesis is a very early event in cell proliferation and takes place before any increase in protein or nucleic acid synthesis. Thus, the polyamine biosynthetic pathway presents an attractive target in tumor and proliferative disease chemotherapy. Inhibitors of polyamine biosynthesis have been used successfully in the treatment of some protozoan diseases, notably African trypanosomiasis.

Adequate intracellular levels of polyamines are necessary for optimal growth and replication of animals, plants, bacteria, fungi, protozoa, and probably all living organisms. Polyamines influence the transcriptional and translational stages of protein synthesis, interact with nucleic acids, stabilize membranes, alter intracellular free calcium, and have important neurophysiological functions. Fruit-ripening and flower-setting processes in plants are modulated by polyamines.

From the foregoing it is apparent that polyamines are, or should be, of interest to plant and animal biochemists and physiologists, cell biologists, microbiologists, neurophysiologists, pharmacologists, and oncologists, to name but a few! It is hoped that *Polyamine Protocols*, which brings together clear and complete descriptions of methods currently in use in the polyamine field, will be of help to workers in these and other disciplines. Every chapter has been written by authors with "hands-on" experience with the techniques they describe. The result is a mixture of "state-of-the-art" procedures together with some older, less sophisticated, and less expensive (!) methods that are still capable of producing reliable data. I thank all the authors for taking the time to contribute, and for prompt delivery of their manuscripts.

Acknowledgment

Work from the author's laboratory was supported by The British Heart Foundation.

David M. L. Morgan

Contents

Contributors

JIM BLANKENSHIP • *Department of Physiology and Pharmacology, School of Pharmacy, University of the Pacific, Stockton, CA*

CATHERINE S. COLEMAN • *Department of Cellular and Molecular Physiology, Milton S. Hershey Medical Center, Pennsylvania State University, Hershey, PA*

CHRISTOPHER P. DENTON • *Vascular Biology Research Center, King's College, London, UK*

DEBORAH M. EVANS • *Departments of Medicine and Therapeutics and Biomedical Sciences, University of Aberdeen, UK*

ALAN H. FAIRLAMB • *Department of Medical Parasitology, London School of Hygiene and Tropical Medicine, London, UK*

ANTONIO FERRANTE • *Department of Immunology, Women's and Children's Hospital, North Adelaide, Australia*

KARL J. HUNTER • *Biotransformation Unit, Unilever Research, Sharnbrook, UK*

SARAH A. LE QUESNE • *Department of Medical Parasitology, London School of Hygiene and Tropical Medicine, London, UK*

A. JILL MACKAREL • *Department of Medicine and Therapeutics, University College, Dublin, Ireland*

RENTALA MADHUBALA • *School of Life Sciences, Jawaharlal Nehru University, New Delhi, India*

DAVID M. L. MORGAN • *Vascular Biology Research Center, King's College, London, UK*

ANTHONY E. PEGG • *Department of Cellular and Molecular Physiology, Milton S. Hershey Medical Center, Pennsylvania State University, Hershey, PA*

LISA M. SHANTZ • *Department of Cellular and Molecular Physiology, Milton S. Hershey Medical Center, Pennsylvania State University, Hershey, PA*

R. JAMES STORER • *Headache Group, Institute of Neurology, London, UK*

AMALIA TABIB • *Department of Molecular Biology, Hebrew University-Hadassah Medical School, Jerusalem, Israel*

HEATHER M. WALLACE • *Departments of Medicine and Therapeutics and Biomedical Sciences, University of Aberdeen, UK*

LAURIE WIEST • *Department of Cellular and Molecular Physiology, Milton S. Hershey Medical Center, Pennsylvania State University, Hershey, PA*

I

INTRODUCTION

1

Polyamines

An Introduction

David M. L. Morgan

1. Introduction

Although receiving little attention in biochemical and physiological textbooks, the polyamines have a long history and have accumulated a considerable literature (for reviews *see* **refs.** ***1–11***). In 1678 Antoni van Leeuwenhoek ***(12)*** described crystals that formed in samples of human semen that had been left to cool. It is now clear that these crystals were spermine phosphate. The phenomenon was rediscovered several times during the next 200 years, in each case the discoverer apparently unaware of what had gone before. By the end of the 19th century there were ten different names for these crystals.

The structures of the three most commonly occurring natural polyamines, putrescine, spermidine, and spermine, together with those of 1,3-diaminopropane and cadaverine (1,5-diaminopentane), are shown in **Fig. 1**. One or more of these compounds are present in every living cell. All have been found in eukaryotes, but spermine rarely occurs in prokaryotes. In addition to spermine, spermidine, and putrescine, a large number of other linear, and some branched-chain, polyamines have been detected in mammalian tissues and excreta, or in plants, bacteria, and microorganisms ***(13–16)*** **(Table 1)**. The polymer-like nature of these compounds has led to an abbreviated way of denoting their structures. Because the molecule starts with a primary amino group (H_2N-) linked to a chain of methylene groups (-CH_2-) interspersed with secondary amino groups (-NH-), and terminates with another primary amino group (-NH_2); the structure can be described simply by denoting the number of carbons between each amino group. Thus putrescine becomes **4**, spermidine **3-4**, spermine **3-4-3**, and so on **(Fig. 1** and **Table 1)**. Branched-chains are denoted by

From: *Methods in Molecular Biology, Vol. 79: Polyamine Protocols*
Edited by: D. Morgan Humana Press Inc., Totowa, NJ

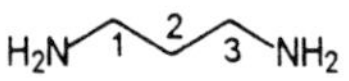

Diaminopropane
(3)

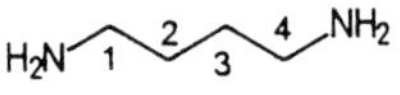

Putrescine
(4)

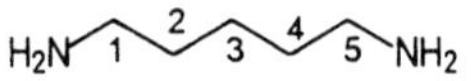

Cadaverine
(5)

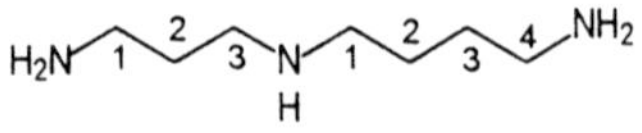

Spermidine
(3-4)

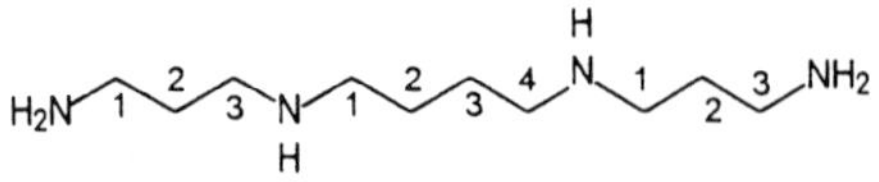

Spermine
(3-4-3)

Fig. 1. Structures of putrescine, spermidine, and spermine, together with diaminopropane and cadaverine. Figures in parentheses below the name correspond to the numbering system described in **Subheading 1.** and are derived from the smaller figures above each bond.

enclosing the side-chain in brackets and placing it next to the atom to which it is attached. The *N*-acetyl polyamines **(Fig. 2)** are the most common naturally occurring derivatives in humans; however, many other polyamine metabolites have also been identified **(Table 2)**.

Polyamines appear to be constituents of many compounds found in plants and insects. *N*-Carbamoylputrescine **(Fig. 3)** and conjugates of hydroxycinnamoyl putrescine, and caffeoylputrescine **(Fig. 3)** are among polyamine intermediates that have been isolated from plant tissues. *N*-Methylputrescine is a precursor of the tropane alkaloids ***(17,18)*** and putrescine-containing alkaloids have also been isolated from the marine gastropod mollusk *Monodonta labio* ***(19)***.

Table 1
Some of the Less Common Polyamines, and the Short Notation for Each Structure[a]

Name	Chemical formula	Abbreviation
Diaminopropane	$NH_2(CH_2)_3NH_2$	3
Putrescine	$NH_2(CH_2)_4NH_2$	4
Cadaverine	$NH_2(CH_2)_5NH_2$	5
Norspermidine	$NH_2(CH_2)_3NH(CH_2)_3NH_2$	3-3
Spermidine	$NH_2(CH_2)_3NH(CH_2)_4NH_2$	3-4
Aminopropylcadaverine	$NH_2(CH_2)_3NH(CH_2)_5NH_2$	3-5
Homospermidine	$NH_2(CH_2)_4NH(CH_2)_4NH_2$	4-4
Norspermine	$NH_2(CH_2)_3NH(CH_2)_3NH(CH_2)_3NH_2$	3-3-3
Thermospermine	$NH_2(CH_2)_3NH(CH_2)_3NH(CH_2)_4NH_2$	3-3-4
Aminopentylnorspermidine	$NH_2(CH_2)_3NH(CH_2)_3NH(CH_2)_5NH_2$	3-3-5
Spermine	$NH_2(CH_2)_3NH(CH_2)_4NH(CH_2)_3NH_2$	3-4-3
Bis(aminopropyl)cadaverine	$NH_2(CH_2)_3NH(CH_2)_5NH(CH_2)_3NH_2$	3-5-3
Aminopropylhomospermine	$NH_2(CH_2)_3NH(CH_2)_4NH(CH_2)_4NH_2$	3-4-4
Canavalmine	$NH_2(CH_2)_4NH(CH_2)_3NH(CH_2)_4NH_2$	4-3-4
Homospermine	$NH_2(CH_2)_4NH(CH_2)_4NH(CH_2)_4NH_2$	4-4-4
Caldopentamine	$NH_2(CH_2)_3NH(CH_2)_3NH(CH_2)_3NH(CH_2)_3NH_2$	3-3-3-3
Aminopropylcanavalmine	$NH_2(CH_2)_3NH(CH_2)_4NH(CH_2)_3NH(CH_2)_4NH_2$	3-4-3-4
Bis(aminopropyl)homospermidine	$NH_2(CH_2)_3NH(CH_2)_4NH(CH_2)_4NH(CH_2)_4NH_2$	3-4-4-3
Bis(aminobutyl)norspermidine	$NH_2(CH_2)_4NH(CH_2)_3NH(CH_2)_3NH(CH_2)_4NH_2$	4-3-3-4
Aminobutylcanavalmine	$NH_2(CH_2)_4NH(CH_2)_3NH(CH_2)_4NH(CH_2)_4NH_2$	4-3-4-4
Aminopropylhomospermine	$NH_2(CH_2)_3NH(CH_2)_4NH(CH_2)_4NH(CH_2)_4NH_2$	3-4-4-4
Homopentamine	$NH_2(CH_2)_4NH(CH_2)_4NH(CH_2)_3NH(CH_2)_4NH_2$	4-4-4-4
N^5-aminobutylhomospermine	$NH_2(CH_2)_3N((CH_2)_4NH_2)(CH_2)_4NH(CH_2)_4NH_2$	4(4)-4-4
Caldohexamine	$NH_2(CH_2)_3NH(CH_2)_3NH(CH_2)_3NH(CH_2)_3NH_2(CH_2)_3NH_2$	3-3-3-3-3
Homocaldohexamine	$NH_2(CH_2)_3NH(CH_2)_3NH(CH_2)_3NH(CH_2)_3NH_2(CH_2)_4NH_2$	3-3-3-3-4
Thermohexamine	$NH_2(CH_2)_3NH(CH_2)_3NH(CH_2)_3NH(CH_2)_4NH_2(CH_2)_3NH_2$	3-3-3-4-3
Homothermohexamine	$NH_2(CH_2)_3NH(CH_2)_3NH(CH_2)_4NH(CH_2)_3NH_2(CH_2)_3NH_2$	3-3-4-3-3
Agmatine	$NH_2C(NH)NH(CH_2)_4NH_2$	
N^6-methylagmatine	$NH_2C(NCH3)NH(CH_2)_4NH_2$	

[a]Also a number of branched-chain penta-, hexa-, and hepta-amines (4(4)-4-4-4; 4-4(4)-4-4-4; 4-4(4)-4-4; 4(4)-4(4)-4; and so on) have been found in plant seeds. Boldface indicates the three most common polyamines.

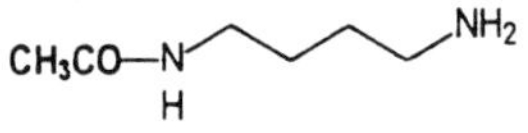

N-Acetylputrescine

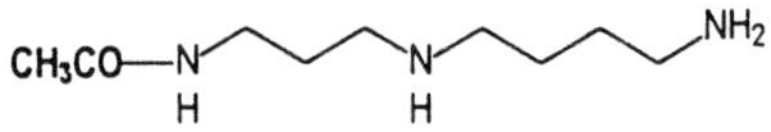

*N*1-Acetylspermidine

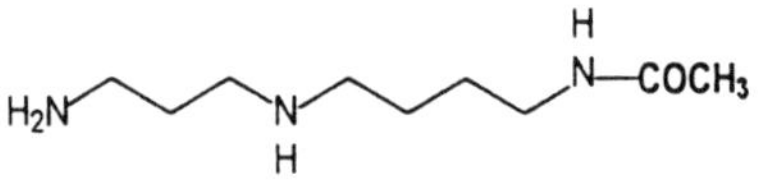

*N*8-Acetylspermidine

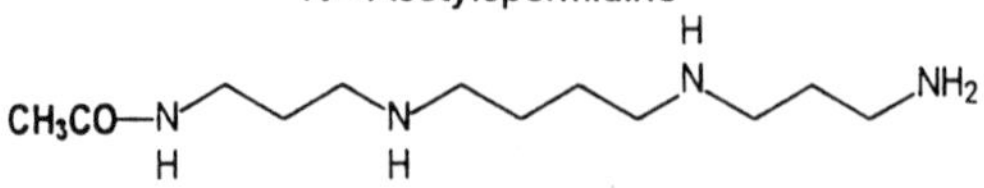

*N*1-Acetylspermine

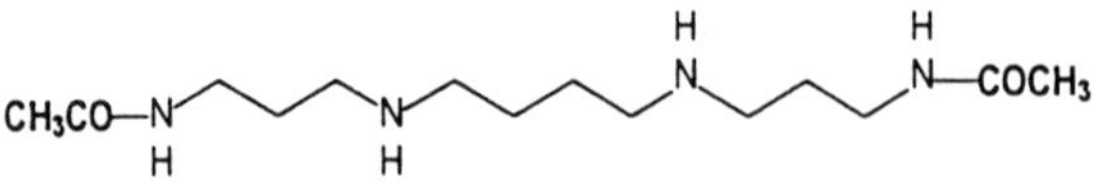

*N*1,*N*12-Diacetylspermine

Fig. 2. Structures of the common *N*-acetyl polyamines (the acetyl groups are in bold type). The numbering system used to indicate the position of the acetyl group is that of Tabor *(23)*.

Palustrine, maytenine, and cannabisativine (from marijuana) are examples of the many spermidine-containing alkaloids found in plants **(Fig. 3)** ***(20)***. Aphelandrine is a spermine-containing alkaloid from the flowering shrub *Aphelandra tetragona*. Low-mol-wt spider and wasp toxins, which are selective inhibitors of glutamate receptors of the central nervous system (CNS), consist of a polyamine backbone to which are linked one or several carboxylic acids and/or amino acids **(Fig. 3)** ***(21)***. Hydroxylamine-containing polyamines have been isolated from the venom of the funnel-web spider *Agelenopsis aperta* ***(22)***. However, in many cases it is not clear whether the polyamines are precursors in the biosynthesis of these compounds. A glutathionyl-spermidine conjugate, N^1-monoglutathionyspermidine is found in *Escherichia coli*; trypanothione,

Table 2
Some Polyamine Metabolites Identified in Human and Rat Urine *(24)*

Diaminopropane
N-Acetyldiaminopropane
β-Alanine[a]
γ-Aminobutyric acid[a]
Cadaverine
N-Acetylcadaverine
δ-Aminovaleric acid
Isoputreanine (*N*-(3-aminopropyl)-4-aminobutyric acid
Putreanine (*N*-(4-aminobutyl)-3-aminopropionic acid
Spermidic acid (*N*-(2-carboxyethyl)-4-aminobutyric acid
Spermic acid-1 *N*-(3-aminopropyl)-*N'*(-(2-carboxyethyl)-1,4-diaminobutane)
Spermic acid-2 (*N,N'*-bis(2-carboxyethyl)-1,4-diaminobutane)

[a]Can also arise from nonpolyamine sources. Polyamine origin of these metabolites was confirmed by the use of isotopically labeled precursors.

a glutathione-spermidine conjugate **(Fig. 4)**, is apparently unique to trypanosomes, where it appears to substitute for glutathione ***(25)***. Castanospermine ***(26)***, aspidospermidine, and aspidospermine ***(27,28)***, are **not** polyamine derivatives.

The polyamines are, in the main, linear aliphatic molecules of, in biological terms, small molecular mass. They are water soluble, and at physiological pH all the amino groups will be positively charged; hence, these compounds are organic bases, their basicity increasing with the number of amino groups. Unlike inorganic molecules or ions, the positive charges on polyamines are spaced out at intervals and, although the hydrocarbon chains are flexible, will have steric as well as cationic properties.

The difficulties experienced in attempting to identify and measure the amounts of polyamines in biological materials follow from three characteristics: their size, the low concentrations in which they are present, and that their only reactive centers are the amino groups. Thus, methods for polyamine analysis must include procedures to extract, and perhaps concentrate, the polyamines; to separate them from other amino-containing compounds (such as amino acids), and from each other; to convert them into colored or fluorescent derivatives; to identify them; and then to make quantitative measurements. Almost all of the methods available to analytical biochemists have been applied to polyamine analysis at some time ***(29,30)*** (*see* Chapters 13–16).

2. Biosynthesis (*see* Chapters 2–5)

With the exception of laboratory-bred mutant cell lines, all cells have the ability to synthesis putrescine and spermidine. Spermine synthesis appears to

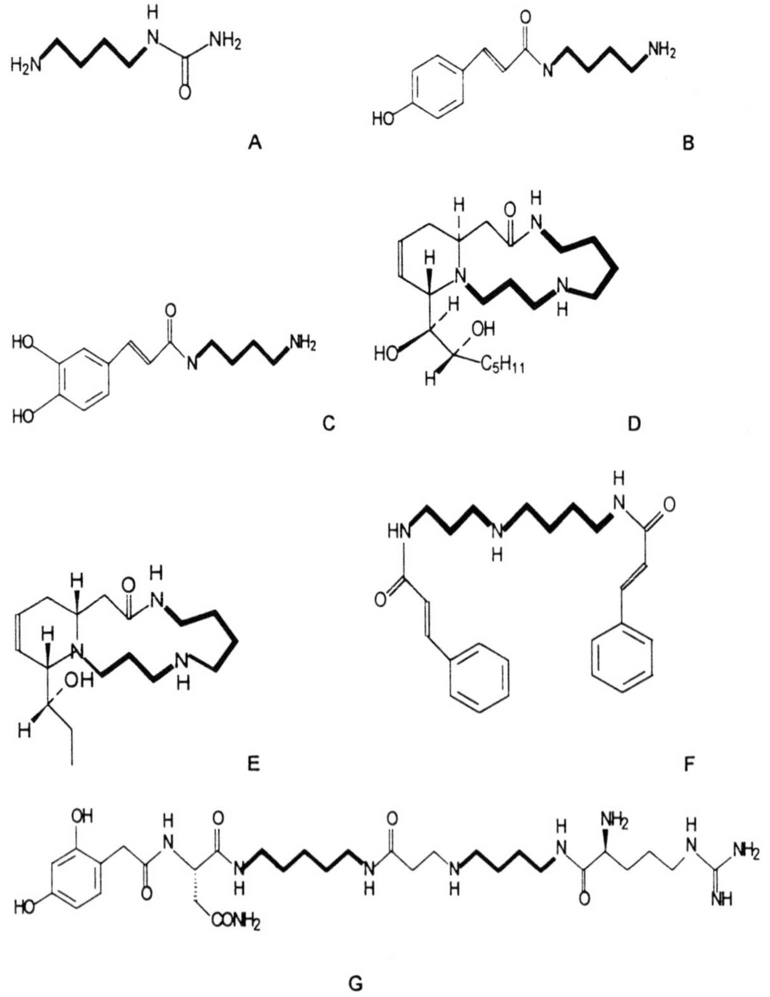

Fig. 3. Examples of polyamine-containing natural compounds: *N*-carbamoylputrescine **(A)**, 4-coumaroylputrescine **(B)**, caffeoylputrescine **(C)**; the spermidine alkaloids cannabisativine **(D)**, palustrine **(E)**, and maytenine **(F)**; the spider toxin NSTX-3 contains both putrescine and cadaverine residues **(G)**. The emphasized bonds indicate the polyamine moieties.

be confined to eukaryotic systems. There are, however, differences of detail between the pathways in different cell types, and a generalized pathway of polyamine biosynthesis is shown in **Fig. 5**. In mammalian cells and in fungi the initial, and at this stage rate-limiting, step is the decarboxylation of ornithine to form putrescine, catalyzed by ornithine decarboxylase. The source of the starting material (ornithine) is not entirely clear. In animals ornithine is present in

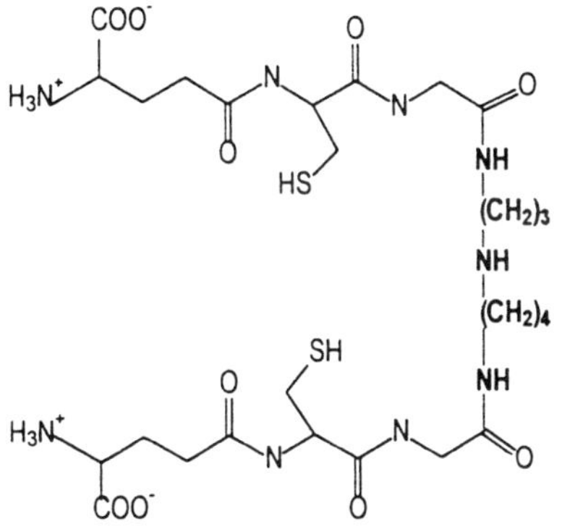

Trypanothione

Fig. 4. The unique glutathione-spermidine conjugate of trypanosomes (the spermidine moiety is in bold type).

the blood plasma (human plasma contains about 85 μmol/L *[31]*), some of which will be dietary in origin. Ornithine is also a product of the urea cycle, and it seems probable that some may be diverted to polyamine biosynthesis. Furthermore, many cells that lack a complete urea cycle contain arginase, and the presence of this enzyme will ensure the availability of ornithine for polyamine production. Thus, in these cells hydrolytic cleavage of the guanidino group of arginine may be the first step in polyamine biosynthesis *(32)*. In mammalian cells and fungi there is only one pathway for putrescine synthesis, but many micro-organisms *(33)* and higher plants *(34,35)* possess a second constitutive pathway via agmatine, itself formed by the decarboxylation of arginine by arginine decarboxylase. Agmatine is then hydrolyzed by agmatinase (agmatine amidinohydrolase) to form putrescine, with the elimination of urea **(Fig. 5)**. Some organisms, e.g., *E. coli*, possess both of these pathways. In plants additional routes exist from agmatine to putrescine; agmatine iminohydrolase yields ammonia and *N*-carbamoylputrescine, which is hydrolyzed by *N*-carbamoylputrescine-amidohydrolase to give ammonia, carbon dioxide, and putrescine **(Fig. 5)**. It is intriguing that agmatine has recently been found in rat tissues *(36)*.

In a step common to most organisms, spermidine is formed from putrescine by addition of an aminopropyl group donated by decarboxylated *S*-adenosylmethionine, a reaction catalyzed by spermidine synthase, an aminopropyltransferase. Addition of a second aminopropyl moiety to spermidine, catalyzed by a different aminopropyl transferase, spermine synthase, forms spermine *(37)*. The source of the aminopropyl group is a second molecule of decarboxylated *S*-adenosylmethionine. The synthesis of spermidine and spermine is

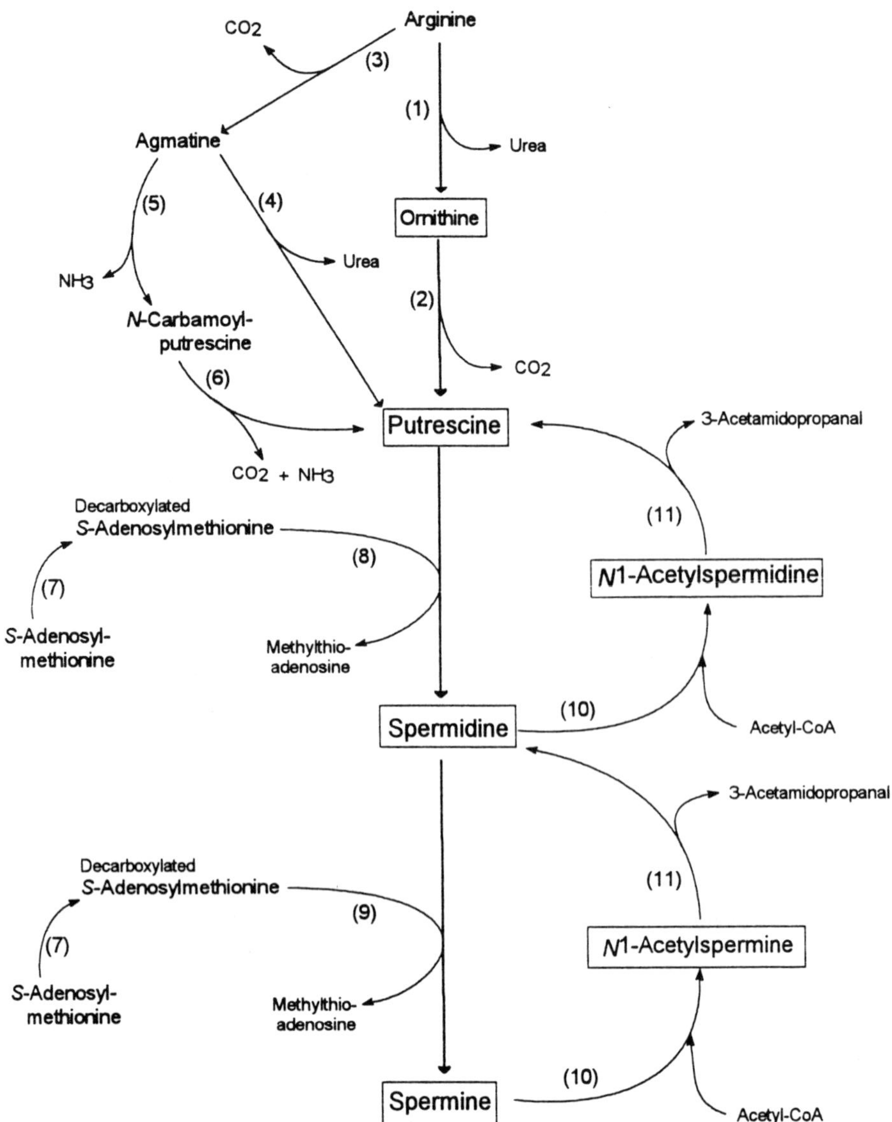

Fig. 5. The main pathways of polyamine biosynthesis in animals, plants, and microorganisms. The enzymes involved are (1) arginase (EC 3.5.3.1); (2) ornithine decarboxylase (EC 4.1.1.17); (3) arginine decarboxylase (EC 4.1.1.19); (4) agmatinase (EC 3.5.3.11); (5) agmatine deiminase (EC 3.5.3.12); (6) *N*-carbamoylputrescine amidase (EC 3.5.1.53); (7) spermidine synthase (EC 2.5.1.16); (8) *S*-adenosylmethionine decarboxylase (EC 4.1.1.50); (9) spermine synthase (EC 2.5.1.22); (10) spermidine/spermine N^1-acetyltransferase (EC 2.3.1.57); (11) polyamine oxidase (EC 1.5.3.11).

dependent on the availability of the aminopropyl donor, hence *S*-adenosylmethionine decarboxylase is also rate-limiting in polyamine biosynthesis. The methionine and adenosine moieties of 5'-methylthioadenosine, the other product of the aminopropyltransferase reactions, are salvaged by a series of reactions that differ between mammalian cells on the one hand, and plants and bacteria on the other, and are not yet fully understood ***(9)***. In a number of bacteria, and in some plants, L-aspartic-β-semialdehyde ***(38)*** is the aminopropyl group donor in spermidine biosynthesis.

The three key enzymes that regulate polyamine biosynthesis are ornithine decarboxylase, *S*-adenosylmethionine decarboxylase, and spermidine/spermine N^1-acetyltransferase ***(39)***, the enzyme that initiates polyamine catabolism; the activities of the other enzymes in the pathway shown in **Fig. 5** appear to be governed by the availability of the appropriate substrate. Each of these enzymes will now be discussed in more detail.

2.1. Ornithine Decarboxylase (E.C.4.1.1.17; see Chapters 2 and 3)

Ornithine decarboxylase (ODC) has received more attention than any other enzyme of the polyamine biosynthetic pathway ***(40–44)***. The enzyme proved difficult to purify because it is extremely labile, the cellular content is very low, and it may be present in multiple forms. However, cDNA clones containing the coding region for ODC have been obtained from a number of species ***(43)***.

All forms of ODC consist of identical subunits and the complete enzyme is generally a dimer. The subunit size of the sequenced enzymes is, in the main, remarkably similar, except for the bacterial and *Leishmania* enzymes, which are about 40% bigger. There are also remarkable similarities in the sequences of the eukaryotic enzymes, with more than 90% identity between the mammalian enzymes. Even *Leishmania* ODC shares 40% identity with mouse ODC over the core region. In contrast, there is very little similarity between the sequences for the bacterial enzymes (from *E. coli*) and those from eukaryotic species. The close identity between the amino acid sequences of the mammalian enzymes suggests that their structural and catalytic properties are also very similar. *E. coli* appears to be unique in that some strains contain two different forms of ODC that are immunologically distinct, a biosynthetic (constitutive) enzyme that resembles those of other organisms and a biodegradative form that may use available ornithine as a carbon source ***(33,45)***.

ODC, from whatever source (including plants), has an absolute requirement for pyridoxal 5'-phosphate for activity. The binding site for pyridoxal phosphate is at lysine-69 in mouse ODC ***(46)***, which is in a highly conserved region within the eukaryotic enzymes. In contrast, the sequence containing the binding site in *E. coli* ODC is completely different ***(43)***. Many forms of the enzyme require the presence of sulfhydryl reducing agents (such as dithiothreitol) for maximal activity and this may well be because of the presence of a cysteine residue in the pyridoxal phosphate binding site.

ODC protein has an extremely short half-life, of the order of 10–60 min, depending on species. Deletion of some or all of the 36 residues at the carboxyl end of the mouse enzyme does not alter enzyme activity, but greatly increases the stability of the protein and prevents its rapid turnover ***(47,48)***. The carboxy terminal part of mouse ODC contains a PEST region (rich in proline, glutamic acid, serine, and threonine; residues 423–449) ***(49)***. *Trypanosoma* ODC does not have a carboxyl PEST sequence and is not rapidly degraded, but this stability was reduced when *Trypanosoma brucei* ODC expressed in Chinese hamster ovary cell had an added mouse ornithine decarboxylase C-terminus ***(50)***. However, *Xenopus* ODC does not have a clear carboxy PEST region, yet has a rapid turnover.

The sole function of ODC appears to be to catalyze the decarboxylation of ornithine, an amino acid not incorporated into proteins, to form putrescine, the initial diamine in the polyamine biosynthetic pathway; 90% of the activity is cytoplasmic. In normal cells ornithine decarboxylase activity is very low and it has been estimated that there may be only 100–200 molecules of enzyme in a quiescent cell ***(51)***.

2.2. S-Adenosylmethionine Decarboxylase (E.C. 4.1.1.50; see Chapter 4)

S-Adenosylmethionine decarboxylase (AdoMetDC) the second of the rate-limiting enzymes in polyamine biosynthesis, is essential for the formation of spermidine and spermidine in mammalian cells ***(52,53)***, plants ***(35,54)***, in micro-organisms such as *E. coli*, *Saccharomyces cerevisiae*, *Physarum polycephalum*, *Neurospora crassa*, and *Aspergillus nidulans* ***(33)*** and probably in all organisms that synthesize these polyamines. All forms of AdoMetDC examined so far are synthesized as a proenzyme. cDNA clones coding for the proenzyme have been isolated from *E. coli*, *S. cerevisiae*, human, bovine, rat, and hamster tissues (*see* **ref.** ***53***). Mammalian AdoMetDC is synthesized as a proenzyme of about 38,000 molecular mass, which is cleaved autocatalytically to form two unequal sized subunits of about 31,000 (α-) and 7000 (β-). The cleavage occurs between glutamate-67 and serine-68; the glutamate becoming the carboxy terminus of the β-subunit whereas the serine becomes the amino terminus of the α-subunit. Both subunits are necessary for catalytic activity; the proenzyme is inactive. The amino-acid sequence of the protein is highly conserved among different mammalian species with about 90% identity among human, rat, bovine, and hamster enzymes.

All known forms of AdoMetDC contain covalently bound pyruvate. On cleavage of the proenzyme, the amino terminal serine of the α-subunit is converted to the pyruvate prosthetic group. In AdoMetDC the carbonyl group of pyruvate is used to form a Schiff base with the substrate, in contrast to the use of pyridoxal phosphate in ornithine decarboxylase.

AdoMetDC is an essential component of the pathway for spermidine and spermine biosynthesis. Decarboxylation of *S*-adenosylmethionine commits this compound to the polyamine pathway and prevents it acting in its usual role as a methyl donor. The active site of AdoMetDC appears to contain a cysteine residue that is essential for enzyme activity, but does not form part of a disulfide bridge. This may explain the stimulatory effect of reducing agents, such as dithiothreitol, on enzyme activity. AdoMetDC has a short half-life of the order of 1 h or less in mammalian cells ***(53)***. There is a PEST region in the α-subunit, but whether this has an effect on the stability of the enzyme has not been established. Breakdown is enhanced in the presence of spermidine or spermine, but not putrescine.

2.3. Spermidine Synthase (E.C. 2.5.1.16) and Spermine Synthase (E.C. 2.5.1.22; see Chapter 5)

Spermidine synthase and spermine synthase are constitutively expressed aminopropyltransferases that are much more stable than either ODC or AdoMetDC. They catalyze the transfer of an aminopropyl group from decarboxylated *S*-adenosylmethionine to putrescine or spermidine; the activities of these synthases are regulated by the availability of their substrates. As a consequence of their lack of a regulatory, or rate-limiting, role in polyamine biosynthesis, these enzymes have been less studied than the two decarboxylases.

Spermidine synthase has been isolated and characterized from bacterial ***(55)***, plant ***(56)***, and mammalian ***(57–61)*** sources. In each case the enzyme consists of two identical subunits of equal size. Spermidine synthase is inhibited by *N*-ethylmaleimide and *p*-hydroxymercuribenzoate and is stimulated by dithiothreitol or 2-mercaptoethanol. Hence, in common with the two decarboxylases, this appears to be a sulfhydryl-requiring enzyme. Cadaverine (1,5-diaminopropane) and spermidine can also act as substrates for the mammalian enzyme, but the reaction proceeds at only 1/20 of the rate with putrescine. Spermidine synthase from bovine brain can also utilize 1,6-diaminohexane as an aminopropyl receptor but only at 1% of the rate with putrescine. Both the reaction products, spermidine and 5'-methylthioadenosine, will inhibit the enzyme. Human spermidine synthase has been cloned ***(62)*** and the amino-acid sequence deduced from the cDNA.

Spermine synthase has been isolated from bovine ***(58)*** and human tissues ***(61)***. Both forms of the enzyme, like the spermidine synthase, consisted of two subunits of equal size. Both were inhibited by the reaction products, spermine and 5'-methylthioadenosine. Putrescine is a competitive inhibitor of spermidine for the bovine enzyme. Spermine synthase is inhibited by *N*-ethylmaleimide and *p*-hydroxymercuribenzoate, and this inhibition can be restored by 2-mercapto-ethanol or dithiothreitol.

3. Polyamine Catabolism (*see* Chapters 6–12)

In contrast to the extensive studies of polyamine biosynthesis, polyamine degradation has received much less attention, possibly because it offers fewer possibilities for control of cellular activity. Polyamines are oxidized in plants, bacteria, fungi, and animals by a variety of oxidases with differing modes of action and cofactor requirements. In mammalian cells spermine and spermidine can be converted back to putrescine by the pathways shown in **Fig. 5.** The first step is the acetylation of one of the two aminopropyl groups of spermine, catalyzed by the enzyme spermidine/spermine N^1-acetyltransferase, to give N^1-acetyl-spermine. This in turn is degraded by a polyamine oxidase with the formation of spermidine and an aldehyde, 3-acetamidopropanal. The single aminopropyl group of spermidine is also acetylated by spermidine/spermine N^1-acetyl-transferase and the resulting N^1-acetylspermidine can be cleaved by a polyamine oxidase to form putrescine and 3-acetamidopropanal. The putrescine can be either recycled to spermidine and possibly spermine **(Fig. 5)** or further metabolized.

3.1. Spermidine/Spermine N^1*-Acetyltransferase (E.C. 2.3.1.57; Diamine* N*-Acetyltransferase;* see *Chapter 6)*

Spermidine/spermine N^1-acetyltransferase (N^1-SAT), which appears to be ubiquitous in mammalian tissues **(39)**, catalyzes the transfer of an acetyl group from acetyl-coenzyme A to an aminopropyl moiety of spermine or spermidine **(Fig. 5)**. The N^1-acetylspermine or N^1-acetylspermidine is then oxidized by the constitutive intracellular polyamine oxidase, which cleaves the polyamine at a secondary amino nitrogen to release 3-acetamid aminopropanal (*N*-acetylamino-propionaldehyde) **(Fig. 6)**. Tissue polyamine oxidase activity is usually sufficient to ensure that intracellular levels of N^1-acetylspermidine or N^1-acetylspermine are below the limits of detection, thus the rate of polyamine degradation is regulated by the activity of N^1-SAT. It is not present in plants, but a similar enzyme has been found in *E. coli* ***(33)***. *Candida boidinii* ***(63)*** contains an acetyltransferase that forms both N^1- and N^8-acetylspermidine in approximately equal proportions, and will also acetylate putrescine and diaminopropane. N^1-SAT has been purified from rat, human, hamster, and chicken tissues (for references *see* **ref.** ***39***). The purified enzyme is unstable and is particularly sensitive to heat. It is not clear whether spermidine or spermine is the preferred substrate in vivo; both polyamines are acetylated in cells in which the enzyme has been activated. N^1-SAT acetylates the aminopropyl end of spermidine, only

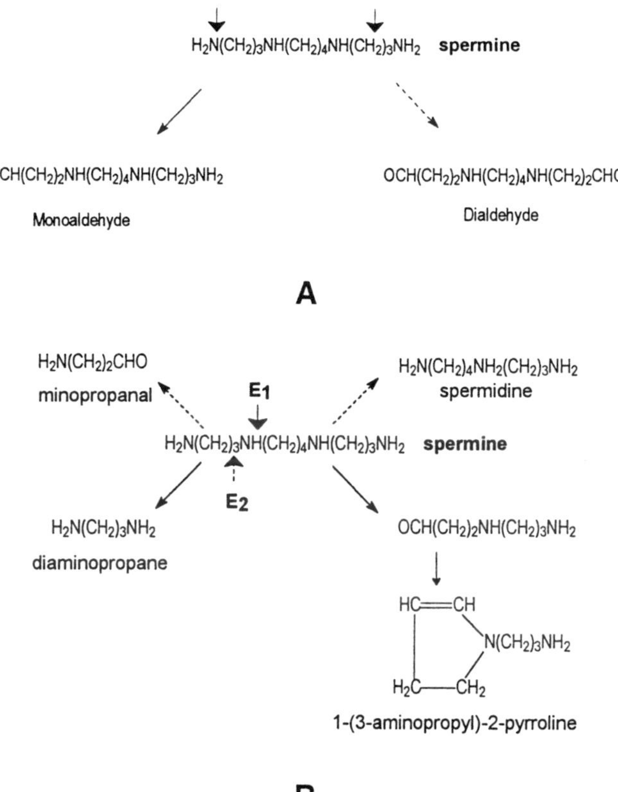

Fig. 6. Alternative sites of action on spermine by polyamine oxidases from different sources. **(A)** Oxidation of terminal (primary) amino groups with the production of ammonia by, e.g., the bovine serum enzyme. The arrows indicate the bond that is broken. **(B)** Enzymes that cleave the molecule at a secondary amino group to produce either diaminopropane (E_1, e.g., maize polyamine oxidase ***[83]*** or aminopropanal E_2, from rat liver ***[84]***).

N^1-acetylspermidine **(Fig. 2)** is produced; N^8-acetylspermidine is not formed. However, other amines containing an aminopropyl group linked to a secondary amino can also act as substrates ***(64)***; but, interestingly, N^8-acetylspermidine is not further acetylated. With spermine the product is N^1-acetylspermine that can be further acetylated to form N^1,N^{12}-diacetylspermine ***(65,66)***, a compound that has been detected in the urine and amniotic fluid of pregnant women ***(67)***. N^1-SAT has no deacetylase activity, hence the catalytic mechanism is essentially irreversible. Human N^1-SAT crossreacts with antisera raised against the rat liver enzyme ***(68)***, indicating a high degree of immunological identity.

The nucleotide sequences of cDNAs have been determined for the human *(69)*, hamster *(70)*, and mouse *(71)* enzymes. The deduced amino acid sequences are more than 95% homologous. Such extensive interspecies homology shows a high degree of evolutionary conservation and is usually taken to indicate that such proteins have important cellular functions.

3.2. Spermidine N^8-Acetyltransferase and Acetylspermidine Deacetylase (see Chapters 7 and 8)

While not part of the main pathway of polyamine metabolism **(Fig. 5)**, it seems appropriate to discuss these two enzymes here. In contrast to spermine/spermidine N^1-acetyltransferase, which is a cytosolic enzyme, spermidine N^8-acetyltransferases (there may be more than one) are predominantly nuclear and can also acetylate histones *(72,73)*. The N^1- and N^8-acetyltransferases are immunologically distinct; antisera to the former having no effect on the latter *(74)*. In contrast to the N^1-acetyltransferase, spermidine N^8-acetyltransferase activity in cultured cells is not affected by serum-stimulated growth or addition of spermidine to the culture medium *(75)*. The N^8-acetylspermidine formed in the nucleus is exported to the cytosol where the majority is deacetylated by a cytosolic acetylspermidine deacetylase *(76,77)* that, contrary to earlier reports, does not act on N^1-acetylspermidine *(78)*. The function of spermidine N^8-acetyltransferase is not clear but, by analogy with what is known at the cellular level, it seems probable that acetylation may be a means of inactivating nuclear spermidine and converting it to a form that can more easily cross the nuclear membrane. The deacetylase then regenerates the spermidine to become part of the intracellular pool.

3.3. Polyamine Oxidases (see Chapters 9–12)

Polyamines are oxidized to form aminoaldehydes by a number of enzymes, classified variously as EC 1.4.3.4. amine oxidase (flavin-containing); EC 1.4.3.6. amine oxidase (copper-containing); EC 1.4.3.10. putrescine oxidase, and EC 1.5.3.11. polyamine oxidase. The trivial names associated with the first two groups by the Nomenclature Committee of the International Union of Biochemistry *(79)* (monoamine oxidase, amine oxidase, tyramine oxidase; and diamine oxidase, amine oxidase, histaminase, respectively) are indicative of the confusion in this area. The putrescine oxidase of *Micrococcus rubens*, a diamine oxidase, contains flavin *(80)*, whereas the diamine oxidase in porcine plasma is a copper-containing enzyme *(81)*. Hence, classification of a particular amine oxidase within these groups is often difficult, particularly so in the case of partially purified preparations. An alternative approach *(5,82)*, is to divide the polyamine oxidases into those that act at the primary (terminal) amino groups of di- and polyamines and form ammonia as one of the products **(Fig.**

6A; Group I); and those (the majority) that act at the secondary amino group(s) of the aminopropyl moieties of spermine or spermidine **(Fig. 6B**; Group II). Enzymes cleaving a polyamine at a secondary amino group would be further subdivided according to whether diaminopropane (Group IIa) or 3-acetamidopropanal (aminopropionaldehyde; Group IIb) are among the products **(Fig. 6)**. However, there are also difficulties with this classification, because it may prove impossible to separate putrescine- from spermidine- or spermine-oxidizing activities, although the latter involve attack at a secondary amino group.

Amine oxidases able to utilize the polyamines spermine or spermidine as substrates will here be considered as polyamine oxidases, whether or not they can also act on di- or mono-amines. Some of the properties of a representative (and incomplete!) selection of amine oxidases are in **Table 3**.

Oxidation taking place at the primary (terminal) amino groups has been named terminal catabolism by Seiler ***(85)***, because the amino aldehydes produced cannot be recycled to polyamines. In contrast, oxidation at a secondary amino group with the formation of putrescine or spermidine that can be recycled back to higher polyamines is part of the interconversion pathway in Seiler's terminology.

3.4.1. Copper-Containing Oxidases

The bovine plasma polyamine oxidase (EC 1.4.3.6.), can be considered the prototypic mammalian copper-dependent amine oxidase. The enzyme acts on spermidine and spermine to produce, respectively, an aminomonoaldehyde [*N*'-(4-aminobutyl)-aminopropanal] or a dialdehyde [*NN*'-*bis*(3-propanal)-1, 4-diaminobutane], ammonia, and hydrogen peroxide **(Fig. 6)** ***(86)***; molecular oxygen is required ***(87)***. The enzyme oxidizes polyamines containing primary amino groups, those forming part of an aminopropyl moiety being the more readily attacked ***(88)***, as well as putrescine ***(89,90)***, and some primary amines. Although hydrogen peroxide is produced, there is no evidence of superoxide formation ***(107)***. Reports that acrolein is a product of the oxidation of spermine or spermidine by this enzyme ***(108–110)*** persist ***(89)***, despite the careful work of Israel et al. ***(111)***, who were unable to detect acrolein formation when synthetic spermine dialdehyde, prepared by unambiguous synthesis, was allowed to break down under physiological conditions.

3.4.2. FAD-Containing Tissue Amine Oxidases

No detailed mechanistic studies have been made of the flavin adenine dinucleotide (FAD)-containing amine oxidases. Peroxisomal and/or cytoplasmic FAD-containing enzymes are found in most mammalian tissues ***(84,112)***. Rat liver polyamine oxidase ***(84)*** catalyzes the oxidation of spermine, and spermidine, but not putrescine. It acts on the secondary amino groups of spermine

Table 3
A Representative Selection of Amine Oxidases

Source	M_r (kDa)	Subunit M_r (kDa)	Cu(II)	Fe(II)	Inhibition by carbonyl reagents	FAD	Inhibition by SH reagents	Reference
Bacteria								
Micrococcus rubens	52					+	+	***80,91***
Fungi								
Aspergillus niger	255	83	3		+			***92,93***
Plants								
Glucine max (soybean)	113	77	2				–	***94,95***
Zea mays (maize)								
(PAO[1])	65				–	+	–	***96***
(DAO[2])	118	55					+	***97***
Avena sativa (oat)	63					+		***98***
Pisum sativum (pea)	131	66						***99***
Nematode								
Ascaris suum	66	66	–	–	–	+	+	***100***
Mammals								
Bovine plasma	180	85	2		+			***101,102***
Bovine liver	60	57				+		***103***
Rat liver	60	60				+	+	***84***
Human pregnancy plasma	67			+				***104***
Human placenta	70	69.5	1[3]					***105***

[a]1: Polyamine oxidase; 2: diamine oxidase; 3: also contains manganese. Inhibition by carbonyl reagents, primarily semicarbazide, is considered to indicate the presence of topaquinone ***(106)***. Sulfhydryl reagents used were mainly *N*-ethylmaleimide, or *p*-hydroxymercuribenzoate.

or spermidine with 3-aminopropanal as one of the products **(Fig. 6)**. This was confirmed by demonstrating the formation of *N*-acetyl-3-aminopropanal, N^1-acetylspermidine, and putrescine from N^1,N^{12}-diacetylspermine by this enzyme ***(113)***. This is the FAD-dependent enzyme of **Fig. 5**, responsible for the recycling of spermine and spermidine to putrescine. An enzyme with broadly similar characteristics, except that putrescine is also oxidized, is present in human serum during pregnancy ***(104,114)***. Polyamine oxidase activity is high in most tissues. It is present at levels comparable to those of spermine and spermidine synthases and that greatly exceed that of spermine/spermidine N^1-acetyltransferase ***(9)***. Hence, under normal conditions tissue levels of acetylated polyamines verge on the undetectable. This FAD-containing polyamine oxidase is present in all mammalian tissues and in all cultured mammalian cells examined so far.

4. Regulation of Intracellular Polyamine Levels

Although intracellular polyamine concentrations vary throughout the cell cycle ***(115)***, the maximum change rarely exceeds twofold. Polyamine concentrations in resting fibroblasts have been calculated as ≈29 μ*M* putrescine, 159 μ*M* spermidine, and 635 μ*M* spermine; transformed and tumor cells generally have higher levels and for SV-3T3 cells the calculated concentrations are 229, 1835, and 694 μ*M*, respectively ***(116)***. Intracellular polyamine levels are tightly regulated by mechanisms that control biosynthesis, degradation, and uptake. ODC activity is increased by growth factors, hormones, regenerative stimuli, tumor promoters, immunoadjuvants, and many drugs ***(42,117,118)***; in most cases this is because of an increase in enzyme protein. Translation and turnover of ODC protein is regulated, both upward and downward, by changes in intracellular polyamine levels. Degradation of the enzyme protein is enhanced by binding to a unique inhibitory protein, antizyme, which is itself induced by polyamines ***(119)***.

The supply of decarboxylated *S*-adenosylmethionine is highly regulated by changes in the amount of active AdoMetDC protein within the cell resulting from alterations in the rate of formation of active enzyme from proenzyme, in the rate of synthesis of proenzyme, and in the rate of degradation of active enzyme. AdoMetDC is inhibited by its product, decarboxylated *S*-adenosylmethionine, and some forms are activated by putrescine, probably by an allosteric mechanism involving a conformational change in the enzyme protein. Low polyamine levels stimulate both synthesis and translation of AdoMetDC mRNA ***(120,121)***; synthesis of AdoMetDC proenzyme is also influenced by intracellular polyamines. Spermidine and spermine synthases are inhibited by their products, spermidine and spermine, and also by 5'-methylthioadenosine.

N^1-SAT is normally present only at very low levels in cells. It is induced by spermidine and spermine, and also by a number of other agents, including hormones, growth factors, and a variety of toxic insults ***(39)***. In common with the other key enzymes of polyamine metabolism, ornithine decarboxylase, and *S*-adenosylmethionine decarboxylase, N^1-SAT is a short-lived enzyme that is rapidly turned over and is regulated at the levels of mRNA accumulation, message translation, and protein turnover.

Many cell types ***(122,123)*** possess an uptake system for polyamines that is distinct from those for amino acids and that, in certain circumstances, can substitute for synthesis *de novo*. In some cells polyamines are taken up by both saturable and nonsaturable systems. Uptake by saturable systems is energy-requiring, temperature-dependent, carrier-mediated, and operates against a substantial concentration gradient. In many cells putrescine and, often, spermidine transport is sodium-sensitive (the term sodium-sensitive is preferable to the more frequently used sodium-dependent, because removal of extracellular sodium reduces, but often does not abolish, uptake). In contrast, spermine transport in most cells is not affected by changes in extracellular sodium. Transport can be inhibited by reagents, such as *N*-ethylmaleimide or *p*-chloromercuribenzoate, that alkylate sulfhydryl groups, and stimulated by sulfhydryl donors, such as dithiothreitol, indicating a requirement for SH-groups on the transporter(s) that may be within, rather than on, the cell membrane. The number of carriers in the polyamine transport system varies with cell type. In human umbilical vein endothelial cells there are apparently two carriers, one shared by spermine and spermidine, and one capable of transporting all three polyamines ***(124)***. In contrast, porcine aortic endothelial cells appear to possess three carriers, one for each polyamine ***(125)***. Multiple carriers are not uncommon, although many cell types appear to possess a single polyamine carrier ***(126)***.

The mechanisms of polyamine transport and the means by which polyamine uptake is induced are not clear. Uptake is generally low in quiescent cells, or in cells that have been induced to differentiate, in contrast to cells in rapid growth, in which uptake is enhanced. Uptake is also increased in response to proliferative stimuli, such as serum, growth factors, and hormones, and in cells in which polyamine stores have been depleted. Polyamine transport appears to be regulated by both a rapidly degraded protein inhibitor, probably antizyme, an ODC inhibitory protein ***(119,127)*** that responds to a rise in intracellular polyamine levels, and a longer-lived protein, or proteins, which may be the polyamine carrier(s).

5. Polyamine Pharmacology

Induction of ODC is a very early event in cell proliferation, the increase in enzyme activity occurring before DNA or protein synthesis. This led to considerable interest in the use of polyamines as tumor markers or as indices of tumor therapy ***(41,128,129)***, but this proved less successful than expected ***(130)***.

Polyamine biosynthesis is also seen as a potential target in tumor chemotherapy ***(8,10,131)*** and in other proliferative diseases. Inhibitors have been developed to block one of the regulatory steps in the pathway ***(39,43,53,131–134)***. Another ongoing approach has been the development of structural analogs of the polyamines that would be taken up by the polyamine transport system and selectively interfere in polyamine metabolism ***(135)***. Inhibitors of polyamine biosynthesis have been used successfully in the treatment of some cancers and protozoan diseases, particularly African trypanosomiasis ***(132,136,137)***.

6. Functions of the Polyamines

What do the polyamines do? The conventional response is to say that at present we do not know, but this has been described as "a cautious misstatement" (Marton and Morris, in **ref. *132***, p. 79). It is clear that adequate intracellular levels of polyamines are necessary for optimal growth and replication of plant, bacteria, fungi, and, indeed, all cell types examined so far. Intracellular concentrations of polyamines are in the high micromolar range and at physiological pH all their amino groups will be positively charged. Hence, the majority of the polyamines will be sequestered in some way and it is probable that only the "free" polyamine pool is physiologically active. Attempts to localize intracellular polyamines have been inconclusive ***(138)*** and, as yet, no polyamine stores have been identified. The first unequivocally established function for polyamines at the molecular level is the donation of a 4-aminobutyl moiety by spermidine to the eukaryotic initiation factor 5A (eIF-5A) precursor protein to form the amino acid hypusine ***(139)***. Polyamines can influence the transcriptional and translational stages of protein synthesis (*see* **ref. *9***), stabilize membranes ***(140)***, alter intracellular free calcium levels ***(141–143)***, and may well have important messenger functions. Polyamines also have important neurophysiological functions ***(144–146)***. To quote Coffino ***(147)***, "Polyamines are doing some important things, but we do not know what they are."

References

1. Cohen, S. S. (1971) *Introduction to the Polyamines.* Prentice-Hall, Englewood Cliffs, NJ.
2. Bachrach, U. (1973) *Function of Naturally Occurring Polyamines.* Academic, New York and London.
3. Gaugas, J. M. (1980) *Polyamines in Biomedical Research.* John Wiley, Chichester.
4. Morris, D. R. and Marton, L. J. (1981) *Polyamines in Biology and Medicine.* Marcel Dekker, New York and Basel.
5. Morgan, D. M. L. (1987) Polyamines. *Essays Biochem.* **23,** 82–115.
6. Tabor, H. and Tabor, C. W. (1983) *Methods Enzymol.* **84,** Academic, New York and London.

7. Bachrach, U. and Heimer, Y. (1989) *Physiology of Polyamines*, 2 vols. CRC, Boca Raton, FL.
8. Pegg, A. E. (1988) Polyamine metabolism and its importance in neoplastic growth and a target for chemotherapy. *Cancer Res.* **48,** 759–774.
9. Pegg, A. E. (1986) Recent advances in the biochemistry of polyamines in eukaryotes. *Biochem. J.* **234,** 249–262.
10. Janne, J., Alhonen, L., and Leinonen, P. (1991) Polyamines: from molecular biology to clinical applications. *Ann. Med.* **23,** 241–259.
11. Slocum, R. D. and Flores, H. E. (1991) *Biochemistry and Physiology of Polyamines in Plants.* CRC, Boca Raton, FL.
12. Leuwenhoek, A. van. (1678) Observationes D Anthonii Leuwenhoek de natis semine genitali animalculis. *Phil. Trans. Roy. Soc. London* **12,** 1040–1043.
13. Hamana, K. and Matsuzaki, S. (1992) Polyamines as a chemotaxonomic marker in bacterial systematics. *Crit. Rev. Microbiol.* **18,** 261–283.
14. Hamana, K., Matsuzaki, S., Niitsu, M., and Samejima, K. (1994) Distribution of unusual polyamines in aquatic plants and gramineous seeds. *Can. J. Bot.* **72,** 1114–1120.
15. Hamana, K., Hamana, H., Niitsu, M., Samejima, K., Sakane, T., and Yokota, A. (1994) Occurrence of tertiary and quaternary branched polyamines in thermophilic archaebacteria. *Microbios* **79,** 109–119.
16. Hamana, K., Hamana, H., Niitsu, M., Samejima, K., Sakane, T., and Yokota, A. (1993) Tertiary and quaternary branched polyamines distributed in thermophilic *Saccharococcus* and *Bacillus. Microbios* **75,** 23–32.
17. Lounasmaa, M. (1988) The tropane alkaloids. *Alkaloids* **33,** 1–81.
18. Leete, E. (1990) Recent developments in the biosynthesis of the tropane alkaloids. *Planta Med.* **56,** 339–352.
19. Niwa, H., Watanabe, M., and Yamada, K. (1993) Monodontamides A, B, and C, three new putrescine alkaloids from the marine gastropod mollusc *Monodonta labio* (Linné). *Tetrahedron Lett.* **34,** 7441–7444.
20. Goggisberg, A. and Hesse, M. (1983) Putrescine, spermidine, spermine, and related polyamine alkaloids. *Alkaloids* **22,** 85–188.
21. Saccomano, N. A., Volkmann, R. A., Jackson, H., and Parks, T. N. (1989) Polyamine spider toxins: unique pharmacological tools. *Ann. Rep. Med. Chem.* **24,** 287–293.
22. Jasys, V. J., Kelbaugh, P. R., Nason, D. M., Phillips, D., Rosnack, K. J., Saccomano, N. A., Stroh, J. G., and Volkman, R. A. (1990) Isolation, structure elucidation, and synthesis of novel hydroxylamine-containing polyamines from the venom of the Agelenopsis aperta spider. *J. Am. Chem. Soc.* **112,** 6696–6704.
23. Tabor, H., Tabor, C. W., and De Meis, L. (1971) Chemical synthesis of *N*-acetyl-1,4-diaminobutane, N^1-acetylspermidine, and N^8-acetylspermidine. *Methods Enzymol.* **17B,** 829–833.
24. Muskiet, F. A. J., Dorhout, B., van den Berg, G. A., and Hessels, J. (1995) Investigation of polyamine metabolism by high-performance liquid chromatographic and gas chromatographic profiling methods. *J. Chromatogr. B Biomed. Appl.* **667,** 189–198.

25. Fairlamb, A. H. and Cerami, A. (1992) Metabolism and functions of trypanothione in the kinetoplastida. *Ann. Rev. Microbiol.* **46,** 695–729.
26. Hohenschutz, L. D., Bell, E. A., Jewess, P. J., Leworthy, D. P., Pryce, R. J., Arnold, E., and Clardy, J. (1981) Castanospermine, a 1,6,7,8-tetrahydroxyocta-hydroindolizine alkaloid, from seeds of *Castanospermum australe. Phytochemistry* **20,** 811–814.
27. Saxton, J. E. (1983) The Aspidosperma group, in *Heterocyclic Compounds, vol. 25, pt. 4. The Monoterpenoid Indole Alkaloids* (Saxton, J. E., ed.), Wiley, New York, pp. 331–437.
28. Cordell, G. A. (1983) The bisindole alkaloids, in *Heterocyclic Compounds, vol. 25, pt. 4. The Monoterpenoid Indole Alkaloids* (Saxton, J. E., ed.), Wiley, New York, pp. 539–728.
29. Seiler, N. (1980) Assay of polyamines in tissues and body fluids, in *Polyamines in Biomedical Research* (Gaugas, J. M., ed.), Wiley, Chichester, pp. 435–461.
30. Seiler, N. (1986) Polyamines. *J. Chromatogr.* **379,** 157–176.
31. Liappis, N. (1972) Geschlechtsspezifische Unterschiede der Freien Aminosauren im Serum von Erdwachsenen. *Z. Klin. Chem. Klin. Biochem.* **10,** 132–135.
32. Pegg, A. E. and McCann, P. P. (1982) Polyamine metabolism and function. *Am. J. Physiol.* **243,** C212–C221.
33. Tabor, C. W. and Tabor, H. (1985) Polyamines in microorganisms. *Microbiol. Rev.* **49,** 81–99.
34. Slocum, R. D., Kaur-Sawhney, R., and Galston, A. (1984) The physiology and biochemistry of polyamines in plants. *Arch. Biochem. Biophys.* **235,** 283–303.
35. Smith, T. A. (1985) Polyamines. *Ann. Rev. Plant Physiol.* **36,** 117–143.
36. Raasch, W., Regunathan, S., Li, G., and Reis, D. J. (1995) Agmatine, the bacterial amine, is widely distributed in mammalian tissues. *Life Sci.* **56,** 2319–2330.
37. Pegg, A. E., Shuttleworth, K., and Hibasami, H. (1981) Specificity of mammalian spermidine synthase and spermine synthase. *Biochem. J.* **197,** 315–320.
38. Tait, G. H. (1985) Bacterial polyamines, structures and biosynthesis. *Biochem. Soc. Trans.* **13,** 316–318.
39. Casero, R. A., Jr. and Pegg, A. E. (1993) Spermidine/spermine N^1-acetyltransferase—The turning point in polyamine metabolism. *FASEB J.* **7,** 653–661.
40. Canellakis, E. S., Viceps-Madore, D., Kyriakidis, D. A., and Heller, J. S. (1979) The regulation and function of ornithine decarboxylase and of the polyamines. *Curr. Topics Cell Regul.* **15,** 155–201.
41. Russell, D. H. (1983) Clinical relevance of polyamines. *CRC Crit. Rev. Clin. Lab. Sci.* **18,** 261–311.
42. Hayashi, S. (ed.) (1989) *Ornithine Decarboxylase: Biology, Enzymology, and Molecular Genetics.* Pergamon, Oxford and New York.
43. McCann, P. P. and Pegg, A. E. (1992) Ornithine decarboxylase as an enzyme target for therapy. *Pharmacol. Ther.* **54,** 195–215.
44. Davis, R. H., Morris, D. R., and Coffino, P. (1992) Sequestered end products and enzyme regulation: the case of ornithine decarboxylase. *Microbiol. Rev.* **56,** 280–290.

45. Applebaum, D. M., Dunlap, J. C., and Morris, D. R. (1977) Comparison of the biosynthetic and biodegradative ornithine decarboxylases of *Escherichia coli. Biochemistry* **16,** 1580.
46. Poulin, R., Lu, L., Ackerman, B., Bey, P., and Pegg, A. E. (1992) Mechanisms of the irreversible inactivation of mouse ornithine decarboxylase by a-difluoromethylornithine. *J. Biol. Chem.* **267,** 150–158.
47. Ghoda, L., van Daalen Wetters, T., Macrae, M., Ascherman, D., and Coffino, P. (1989) Prevention of rapid intracellular degradation of ODC by a carboxyl-terminal truncation. *Science* **243,** 1493–1495.
48. Lu, L., Stanley, B. A., and Pegg, A. E. (1991) Identification of residues in ornithine decarboxylase essential for enzymic activity and for rapid protein turnover. *Biochem J.* **277,** 671–675.
49. Rogers, S., Wells, R., and Rechsteiner, M. (1986) Amino acid sequences common to rapidly degraded proteins: the PEST hypothesis. *Science* **234,** 364–368.
50. Ghoda, L., Phillips, M. A., Bass, K. E., Wang, C. C., and Coffino, P. (1990) Trypanosome ornithine decarboxylase is stable because it lacks sequences found in the carboxy terminus of the mouse enzyme which target the latter for intracellular degradation. *J. Biol. Chem.* **265,** 11,823–11,826.
51. Pegg, A. E., Seeley, J. E., Pösö, H., Della Ragione, F., and Zagon, I. S. (1982) Polyamine biosynthesis and interconversion in rodent tissues. *Fed. Proc.* **41,** 3065–3072.
52. Tabor, C. W. and Tabor, H. (1984) Methionine adenosyltransferase (*S*-adenosylmethionine synthase) and S-adenosylmethionine decarboxylase. *Adv. Enzymol. Relat. Areas Mol. Biol.* **56,** 251–282.
53. Pegg, A. E. and McCann, P. P. (1992) *S*-adenosylmethionine decarboxylase as an enzyme target for therapy. *Pharmacol. Ther.* **56,** 359–377.
54. Slocum, R. D., Kaur Sawhney, R., and Galston, A. W. (1984) The physiology and biochemistry of polyamines in plants. *Arch. Biochem. Biophys.* **235,** 283–303.
55. Bowman, W. H., Tabor, C. W., and Tabor, H. (1973) Spermidine biosynthesis. Purification and properties of propylamine transferase from *Escherichia coli. J. Biol. Chem.* **248,** 2480–2486.
56. Hirasawa, E. and Suzuki, Y. (1983) Biosynthesis of spermidine in maize seedlings. *Phytochemistry* **22,** 103–106.
57. Samejima, K. and Yamanoha, B. (1982) Purification of spermidine synthase from rat ventral prostate by affinity chromatography on immobilised *S*-adenosyl(5')-3-thiopropylamine. *Arch. Biochem. Biophys.* **216,** 213–222.
58. Pajula, R. L., Raina, A., and Eloranta, T. (1979) Polyamine synthesis in mammalian tissues: isolation and characterisation of spermine synthase from bovine brain. *Eur. J. Biochem.* **101,** 619–626.
59. Yamanoha, B., Samejima, K., Nakajima, T., and Yasuhara, T. (1984) Differences between homogenous spermidine synthases isolated from rat and pig liver. *J. Biochem.* **96,** 1273–1281.
60. Raina, A., Hyvonen, T., Eloranta, T., Voutilainen, M., Samejima, K., and Yamanoha, B. (1984) Polyamine synthesis in mammalian tissues. Isolation and

characterization of spermidine synthase from bovine brain. *Biochem. J.* **219,** 991–1000.

61. Kajander, E. O., Kauppinen, L. I., Pajula, R. L., Karkola, K., and Eloranta, T. O. (1989) Purification and partial characterization of human polyamine synthases. *Biochem. J.* **259,** 879–886.
62. Wahlfors, J., Alhonen, L., Kauppinen, L., Hyvonen, T., Janne, J., and Eloranta, T. O. (1990) Human spermidine synthase: cloning and primary structure. *DNA Cell Biol.* **9,** 103.
63. Haywood, G. W. and Large, P. J. (1985) The occurrence, subcellular localization and partial purification of diamine acetyltransferase in the yeast Candida boidinii grown on spermidine or putrescine as sole nitrogen source. *Eur. J. Biochem.* **148,** 277–283.
64. Della Ragione, F. and Pegg, A. E. (1983) Studies of the specificity and kinetics of rat liver spermidine/spermine N^1-acetyltransferase. *Biochem. J.* **213,** 701–706.
65. Della Ragione, F. and Pegg, A. E. (1982) Purification and characterization of spermidine/spermine N1-acetyltransferase from rat liver. *Biochemistry* **21,** 6152–6158.
66. Libby, P. R., Ganis, B., Bergeron, R. J., and Porter, C. W. (1991) Characterization of human spermidine/spermine N1-acetyltransferase purified from cultured melanoma cells. *Arch. Biochem Biophys.* **284,** 238–244.
67. van den Berg, G. A., Kingma, A. W., Visser, G. H., and Muskiet, F. A. (1988) Gestational-age-dependent concentrations of polyamines, their conjugates and metabolites in urine and amniotic fluid. *Br. J. Obstet. Gynaecol.* **95,** 669–675.
68. Casero, R. A. J., Celano, P., Ervin, S. J., Wiest, L., and Pegg, A. E. (1990) High specific induction of spermidine/spermine N1-acetyltransferase in a human large cell lung carcinoma. *Biochem. J.* **270,** 615–620.
69. Xiao, L., Celano, P., Mank, A. R., Pegg, A. E., and Casero, R. A. J. (1991) Characterization of a full-length cDNA which codes for the human spermidine/spermine N1-acetyltransferase. *Biochem. Biophys. Res. Commun.* **179,** 407–415.
70. Pegg, A. E., Stanley, B. A., Wiest, L., and Casero, R. A., Jr. (1992) Nucleotide sequence of hamster spermidine/spermine-N^1-acetyltransferase cDNA. *Biochim. Biophys. Acta Gene Struct. Expression* **1171,** 106–108.
71. Fogel-Petrovic, M., Kramer, D. L., Ganis, B., Casero, R. A., Jr., and Porter, C. W. (1993) Cloning and sequence analysis of the gene and cDNA encoding mouse spermidine/spermine N^1-acetyltransferase—A gene uniquely regulated by polyamines and their analogs. *Biochim. Biophys. Acta Gene Struct. Expression* **1216,** 255–264.
72. Libby, P. R. (1978) Calf liver nuclear *N*-acetyltransferases. *J. Biol. Chem.* **253,** 233–237.
73. Libby, P. R. (1980) Rat liver nuclear *N*-acetyltransferases: separation of two enzymes with both histone and spermidine acetyltransferase activity. *Arch. Biochem. Biophys.* **203,** 384–389.
74. Erwin, B. G., Persson, L., and Pegg, A. E. (1984) Differential inhibition of histone and polyamine acetylases by multisubstrate analogues. *Biochemistry* **23,** 4250–4255.

75. Desiderio, M. A., Mattei, S., Biondi, G., and Colombo, M. P. (1993) Cytosolic and nuclear spermidine acetyltransferases in growing NIH 3T3 fibroblasts stimulated with serum or polyamines: relationship to polyamine-biosynthetic decarboxylases and histone acetyltransferase. *Biochem. J.* **293,** 475–479.
76. Libby, P. R. (1978) Properties of an acetylspermidine deacetylase from rat liver. *Arch. Biochem. Biophys.* **188,** 360–363.
77. Blankenship, J. (1978) Deacetylation of N^8-acetylspermidine by subcellular fractions of rat tissue. *Arch. Biochem. Biophys.* **189,** 20–27.
78. Marchant, P., Manneh, V. A., and Blankenship, J. (1986) N1-acetylspermidine is not a substrate for N-acetylspermidine deacetylase. *Biochim. Biophys. Acta.* **881,** 297–299.
79. Nomenclature Committee. (1992) *Enzyme Nomenclature* 1992, Academic, San Diego.
80. Yamada, H., Adachi, O., and Ogata, K. (1965) Putrescine oxidase, a diamine oxidase requiring flavin adenine dinucleotide. *Agr. Biol. Chem.* **29,** 1148,1149.
81. Falk, M. C., Staton, A. J., and Williams, T. J. (1983) Heterogeneity of pig plasma amine oxidase: molecular and catalytic properties of chromatographically isolated forms. *Biochemistry* **22,** 3746–3751.
82. Federico, R. and Angelini, R. (1991) Polyamine catabolism in plants, in *Biochemistry and Physiology of Polyamines in Plants* (Slocum, R. D. and Flores, H. E., eds.), CRC, Boca Raton, FL, pp. 41–56.
83. Smith, T. A. and Barker, J. H. (1988) The Di- and polyamine oxidases of plants. *Adv. Exp. Med. Biol.* **250,** 573–587.
84. Hölttä, E. (1977) Oxidation of spermidine and spermine in rat liver: purification and properties of polyamine oxidase. *Biochemistry* **16,** 91–100.
85. Seiler, N., Bolkenius, F. N., and Knodgen, B. (1985) The influence of catabolic reactions on polyamine excretion. *Biochem. J.* **225,** 219–226.
86. Tabor, C. W., Tabor, H., and Bachrach, U. (1964) Identification of the aminoaldehydes produced by the oxidation of spermine and spermidine with purified plasma amine oxidase. *J. Biol. Chem.* **239**, 2194–2203.
87. Tabor, C. W., Tabor, H., and Rosenthal, S. M. (1954) Purification of amine oxidase from beef plasma. *J. Biol. Chem.* **208,** 645–661.
88. Israel, M., Rosenfield, J. S., and Modest, E. J. (1964) Analogs of spermine and spermidine. I. Synthesis of polymethylene-polyamines by reduction and cyanoethylated *a,w*-alkylenediamines. *J. Med. Chem.* **7,** 710–716.
89. Gahl, W. A. and Pitot, H. C. (1982) Polyamine degradation in foetal and adult bovine serum. *Biochem. J.* **202,** 603–611.
90. Isobe, K., Tani, Y., Yamada, H., and Hiromi, K. (1980) Determination of polyamines with immobilised beef plasma amine oxidase. *Agric. Biol. Chem.* **44,** 615–619.
91. Ishizuka, H., Horinouchi, S., and Beppu, T. (1993) Putrescine oxidase of *Micrococcus rubens*: primary structure and *Escherichia coli. J. Gen. Microbiol.* **139,** 425–432.

92. Smith, T. A. (1983) Polyamine oxidae (oat seedlings). *Methods Enzymol.* **94,** 311–314.
93. Yamada, H., Adachi, O., and Ogata, K. (1965) Amine oxidases of microorganisms: part IV. Further properties of amine oxidase of *Aspergillus niger. Agr. Biol. Chem.* **29,** 912–917.
94. Vianello, F., Di Paolo, M. L., Stevanato, R., Gasparini, R., and Rigo, A. (1993) Purification and characterization of amine oxidase from soybean seedlings. *Arch. Biochem. Biophys.* **307,** 35–39.
95. Suzuki, Y. (1973) Some properties of the amine oxidase of soybean seedlings. *Plant Cell Physiol.* **14,** 413–417.
96. Suzuki, Y. and Yanagisawa, H. (1980) Purification and properties of maize polyamine oxidase—a flavoprotein. *Plant Cell Physiol.* **21,** 1085–1094.
97. Suzuki, Y. and Hagiwara, M. (1993) Purification and characterization of diamine oxidase from *Zea mays* shoots. *Phytochemistry* **33,** 995–998.
98. Federico, R., Alisi, C., Forlani, F., and Angelini, R. (1989) Purification and characterisation of oat polyamine oxidase. *Phytochemistry* **28,** 2045,2046.
99. Nunokawa, Y., Ishida, N., and Tanaka, S. (1993) Cloning of inducible nitric oxide synthase in rat vascular smooth muscle cells. *Biochem. Biophys. Res. Commun.* **191,** 89–94.
100. Müller, S. and Walter, R. D. (1992) Purification and characterization of polyamine oxidase from *Ascaris suum. Biochem J.* **283,** 75–80.
101. Pettersson, G. (1985) Plasma amine oxidase, in *Structure and Functions of Amine Oxidases* (Mondovi, B., ed.), CRC, Boca Raton, FL, pp. 105–120.
102. Morgan, D. M. L. (1989) Polyamine oxidase and oxidised polyamines, in *Physiology of the Polyamines,* vol. I (Bachrach, U. and Heimer, Y., eds.), CRC, Boca Raton, FL, pp. 203–229.
103. Gasparayan, V. K. and Nalbandyan, R. M. (1990) Purification and fluorescent properties of flavin-containing liver polyamine oxidase. *Biochem. USSR* **55,** 1223–1227.
104. Morgan, D. M. L. (1985) Human pregnancy-associated polyamine oxidase: partial purification and properties. *Biochem. Soc. Trans.* **13,** 351,352.
105. Crabbe, M. J. C., Waight, R. D., Bardsley, W. G., Barker, R. W., Kelley, I., and Knowles, P. F. (1976) Human placental diamine oxidase. Improved purification and characterisation of a copper-and manganese-containing amine oxidase with novel substrate specificity. *Biochem. J.* **155,** 679–687.
106. McIntyre, W. S. and Hartmann, C. (1993) Copper-containing amine oxidases, in *Principles and Applications of Quinoproteins* (Davidson, V. L., ed.), Marcel Dekker, New York, pp. 97–171.
107. Castellano, F. N., He, Z., and Greenaway, F. T. (1993) Hydroxyl radical production in the reactions of copper-containing amine oxidases with substrates. *Biochim. Biophys. Acta Gen. Subj.* **1157,** 162–166.
108. Alarcon, R. A. (1964) Isolation of acrolein from incubated mixtures of spermine with calf serum and its effect on mammalian cells. *Arch. Biochem. Biophys.* **106,** 240–242.

109. Alarcon, R. A. (1970) Acrolein. IV. Evidence for the formation of the cytotoxic aldehyde acrolein from enzymatically oxidised spermine or spermidine. *Arch. Biochem. Biophys.* **137,** 365–372.
110. Kimes, B. W. and Morris, D. R. (1971) Inhibition of nucleic acid and protein synthesis in Escherichia coli by oxidised polyamines and acrolein. *Biochem. Biophys. Acta* **228,** 235–244.
111. Israel, M., Zoll, E. C., Muhammad, N., and Modest, E. J. (1973) Synthesis and antitumor evaluation of the presumed cytotoxic metabolites of spermine and *N,N'*-bis(3-aminopropyl)nonane-1,9-diamine. *J. Med. Chem.* **16,** 1–5.
112. Seiler, N., Bolkenius, F. N., Knodgen, B., and Mamont, P. (1980) Polyamine oxidase in rat tissues. *Biochem. Biophys. Acta* **615,** 480–488.
113. Bolkenius, F. N. and Seiler, N. (1981) Acetylderivatives as intermediates in polyamine catabolism. *Int. J. Biochem.* **13,** 287–292.
114. Morgan, D. M. L. (1985) Polyamine oxidases. *Biochem. Soc. Trans.* **13,** 322–326.
115. Mamont, P. S., Bohlen, P., McCann, P. P., Bey, P., Schuber, F., and Tardif, C. (1976) a-Methylornithine, a potent competitive inhibitor of ornithine decarboxylase, blocks proliferation of rat hepatoma cells in culture. *Proc. Natl. Acad. Sci. USA* **73,** 1626–1630.
116. Morgan, D. M. L. (1990) Polyamines and cellular regulation: perspectives. *Biochem. Soc. Trans.* **18,** 1080–1084.
117. Russell, D. H. (1985) Ornithine decarboxylase: a key regulatory enzyme in normal and neoplastic growth. *Drug Metab. Rev.* **16,** 1–88.
118. Morgan, D. M. L. (1994) Polyamines, arginine and nitric oxide. *Biochem. Soc. Trans.* **22,** 879–883.
119. Hayashi, S. (1995) Antizyme-dependent degradation of ornithine decarboxylase, in *Essays in Biochemistry* (Apps, D. K. and Tipton, K. F., eds.), Portland, London, pp. 37–47.
120. Persson, L., Khomutov, A. R., and Khomutov, R. M. (1989) Feedback regulation of S-adenosylmethionine decarboxylase synthesis. *Biochem. J.* **257,** 929–931.
121. White, M. W., Degnin, C., Hill, J., and Morris, D. R. (1990) Specific regulation by endogenous polyamines of translational initiation of *S*-adenosylmethionine decarboxylase mRNA in Swiss 3T3 fibroblasts. *Biochem. J.* **268,** 657–660.
122. Seiler N. and Dezeure F. (1990) Polyamine transport in mammalian cells. *Int. J. Biochem.* **22,** 211.
123. Khan, N. A., Quemener, V., and Moulinoux, J. P. (1991) Polyamine membrane transport regulation. *Cell Biol. Int. Rep.* **15,** 9–24.
124. Morgan, D. M. L. (1992) Uptake of polyamines by human endothelial cells. Characterization and lack of effect of agonists of endothelial function. *Biochem. J.* **286,** 413–417.
125. Bogle, R. G., Mann, G. E., Pearson, J. D., and Morgan, D. M. L. (1994) Endothelial polyamine uptake: selective stimulation by L-arginine deprivation or polyamine depletion. *Am. J. Physiol. Cell Physiol.* **266,** C776–C783.

126. Seiler N. and Dezeure F. (1990) Polyamine transport in mammalian cells. *Int. J. Biochem.* **22,** 211.
127. Hayashi, S., Murakami, Y., and Matsufuji, S. (1996) Ornithine decarboxylase antizyme: a novel type of regulatory protein. *Trends Biochem. Sci.* **21,** 27–30.
128. Scalabrino, G. and Ferioli, M. E. (1982) Polyamines in mammalian tumours. Part II. *Adv. Cancer Res.* **36,** 1–102.
129. Shipe, J. R., Savory, J., and Wills, M. R. (1981) Polyamines as tumor markers. *CRC Crit. Rev. Clin. Lab. Sci.* **16,** 1–34.
130. Oredsson, S. M. and Marton, L. J. (1982) Polyamines: the elusive cancer markers. *Clin Lab. Med.* **2,** 507–518.
131. Marton, L. J. and Pegg, A. E. (1995) Polyamines as targets for therapeutic intervention. *Ann. Rev. Pharmacol. Toxicol.* **35,** 55–91.
132. McCann, P. P., Pegg, A. E., and Sjoerdsma, A. (1987) *Inhibition of Polyamine Metabolism.* Academic, San Diego.
133. Regenass, U., Mett, H., Stanek, J., Mueller, M., Kramer, D., and Porter, C. W. (1994) CGP 48664, a new *S*-adenosylmethionine decarboxylase inhibitor with broad spectrum antiproliferative and antitumor activity. *Cancer Res.* **54,** 3210–3217.
134. Seiler, N. (1991) Pharmacological properties of the natural polyamines and their depletion by biosynthesis inhibitors as a therapeutic approach. *Progr. Drug Res.* **37,** 107–159.
135. Bergeron, R. J. (1986) Methods for the selective modification of spermidine and its homologues. *Accounts Chem. Res.* **19,** 105–113.
136. Fairlamb, A. H. (1990) Future prospects for the chemotherapy of human trypanosomiasis. 1. Novel approaches to the chemotherapy of trypanosomiasis. *Trans. R. Soc. Trop. Med. Hyg.* **84,** 613–617.
137. Fairlamb, A. H. (1990) Trypanothione metabolism and rational approaches to drug design. *Biochem Soc. Trans.* **18,** 717–720.
138. Sarhan, S. and Seiler, N. (1989) On the subcellular localization of the polyamines. *Biol. Chem. Hoppe-Seyler* **370,** 1279–1284.
139. Park, M. H., Wolff, E. C., and Folk, J. E. (1993) Hypusine: its post-translational formation in eukaryotic initiation factor 5A and its potential role in cellular regulation. *BioFactors* **4,** 95–104.
140. Schuber, F. (1989) Influence of polyamines on membrane functions. *Biochem. J.* **260,** 1–10.
141. Morgan, D. M. L., Coade, S. B., and Pearson, J. D. (1990) Polyamines stimulate calcium uptake by human vascular endothelial cells. *Biochem. Soc. Trans.* **18,** 1223,1224.
142. Groblewski, G. E., Hargittai, P. T., and Seidel, E. R. (1992) Ca^{2+}/calmodulin regulation of putrescine uptake in cultured gastrointestinal epithelial cells. *Am. J. Physiol. Cell Physiol.* **262,** C1356–C1363.
143. Khan, N. A., Sezan, A., Quemener, V., and Moulinoux, J.-P. (1993) Polyamine transport regulation by calcium and calmodulin: role of Ca^{2+}-ATPase. *J. Cell. Physiol.* **157,** 493–501.

144. Williams, K., Romano, C., Dichter, M. A., and Molinoff, P. B. (1991) Modulation of the NMDA receptor by polyamines. *Life Sci.* **48,** 469–498.
145. Williams, K. (1994) Modulation of the *N*-methyl-D-aspartate receptor by polyamines: molecular pharmacology and mechanisms of action. *Biochem. Soc. Trans.* **22,** 884–887.
146. Carter, C. (ed.) (1994) *Neuropharmacology of Polyamines.* Academic, San Diego, CA.
147. Coffino, P. and Poznanski, A. (1991) Killer polyamines? *J. Cell Biochem.* **45,** 54–58.

II

Assay Methods for Enzymes of Polyamine Biosynthesis

2

Determination of Ornithine Decarboxylase Activity Using [^{3}H]Ornithine

Amalia Tabib

1. Introduction

Ornithine decarboxylase (EC 4.1.1.17) (ODC) catalyzes the decarboxylation of ornithine to form putrescine. This reaction is the rate-limiting step in the biosynthesis of the naturally occurring polyamines, which regulate proliferation and which are closely linked to neoplastic growth. Because of the central role played by ODC in various biological systems, attempts have been made to develop sensitive, specific, and simple methods for its assay. These methods can be divided into three categories: assay of ODC mRNA; assay of ODC protein; and assay of ODC activity.

1.1. Assay of ODC mRNA

ODC mRNA species are detected by Northern blots or by *in situ* hybridization histochemistry. In both cases, cDNA probes of ODC can be used *(1)*.

1.2. Assay of ODC Protein

The presence of ODC protein in cellular extracts can be determined by the following immunological methods.

1.2.1. Western Blots

This is a convenient method for the determination of ODC protein. Cellular extracts are separated by SDS-polyacrylamide gel electrophoresis *(2)*, and blots are treated with anti-ODC antibodies. The latter can be obtained commercially (from Euro-Diagnostica AB [Malmo, Sweden] or Sigma [St. Louis, MO]). The amount of ODC protein can then be determined by using a second antibody, based on colorimetric (alkaline phosphatase or peroxidase conjugates) or chemilumines-

From: *Methods in Molecular Biology, Vol. 79: Polyamine Protocols*
Edited by: D. Morgan Humana Press Inc., Totowa, NJ

cence (peroxidase) reactions. Radioactive (iodinated) antibodies can also be used. These reactions are fairly sensitive, but are time-consuming and are not suitable for routine assays. Moreover, the amount of ODC protein does not necessarily reflect ODC activity because regulatory mechanisms may affect the activity of the enzyme.

1.2.2. Immunoprecipitation

This method can be used in conjunction with the Western blot assay. Unlabeled or proteins labeled with radioactive amino acid (or with ^{35}S-sulfate) were extracted from the cells or tissues. The ODC protein is then precipitated by anti-ODC antibodies and protein A-sepharose and separated by gel electrophoresis. Unlabeled ODC protein is detected by a second conjugated antibody (*see* **Subheading 1.2.1.**) and the quantity of ODC protein may thus be determined, whereas the labeled protein can be identified by autoradiograph.

1.2.3. Immunohistochemistry

This method has been widely used to detect ODC protein in cells or tissues. Electron microscopical detections are based on the use of gold *(3)* or electron microscope autoradiography using pulse labeling of cells with α-difluoromethyl[5-^{3}H]ornithine ([^{3}H]-DFMO). DFMO is an irreversible inhibitor of ODC and binds covalently to the active enzyme *(4)*. Regular light microscopy can also be used instead of electron microscopy, and ODC protein can be detected by the use of fluorescent antibodies. Even though these immunohistochemical methods are essentially qualitative, quantitative versions have been recently developed ***(5)*** based on scanning the fluorescence in the preparations with laser beams using a computerized fluorescence microscope (ACAS).

1.3. Assay of ODC Activity

The activity of the enzyme can be determined by measuring the amount of the metabolites formed from ornithine during the decarboxylation reaction. These include the diamine, putrescine, or CO_2. For this purpose nonradioactive (colorimetric, fluorometric) and radiometric methods have been employed.

1.3.1. Nonradiometric Methods

Some of the most commonly used methods for the estimation of putrescine are listed below.

1.3.1.1. Change of pH

This is the earliest and the most convenient method for the detection of ODC activity. It has been used to estimate ODC activity in bacteria, such as in *Escherichia coli*. This qualitative method is routinely used for the identification of bacteria ***(6)***.

1.3.1.2. Paper or Thin-Layer Chromatography (TLC) and Paper Electrophoresis

These methods have been used for the separation of amino acids and amines. Separation by chromatographic methods (paper chromatography *(7)* and TLC *(8)*) is based on the different dissolving abilities of the materials in the solvents, whereas paper electrophoresis *(9)* is mainly based on differences in their charge. Putrescine, thus separated, can be estimated colorimetrically by the ninhydrin reaction. Seiler *(8)* introduced a very sensitive method for the assay of polyamines and diamines. The method is based on the separation of the dansyl-derivatives on silica gel plates by TLC. This TLC method for the assay of polyamines is widely used as an analytical tool (*see* Chapter 16).

1.3.1.3. Ion Exchange Chromatography

Ion-exchange chromatography for the separation and quantitation of putrescine and polyamines was developed by Rosenthal and Tabor *(10)*. Dinitro-fluorobenzene was used for the colorimetric assay of amines in the respective fractions. Subsequently, the more sensitive *o*-phthalaldehyde reagent was used *(11)*. This method, which can be automated using an amino-acid analyzer, requires special equipment and is rather time-consuming.

1.3.1.4. High-Pressure Liquid Chromatography (HPLC)

This method *(12)*, which can be automated, is the most commonly used method for the detection and estimation of putrescine and polyamines (*see* Chapters 13–15). It is relatively fast and sensitive, but does require special equipment.

1.3.2. Radiometric Methods

1.3.2.1. Determination of CO_2 *(13)*

Carbon dioxide is one of the decarboxylation products which cannot be determined by nonisotopic methods. [1-^{14}C]-Ornithine has been used for the determination of the activity of ODC. Essentially, the reaction is carried out in a hermetically sealed system and the radioactive carbon dioxide is trapped in an alkaline (potassium hydroxide or hyamine) solution (*see* Chapter 3). This is a sensitive method but its specificity can be questioned. If the radioactive ornithine is completely metabolized and carbon dioxide is formed by nonspecific oxidation of all of the carbons, then the $^{14}CO_2$ evolved could be the result of reactions other than the decarboxylation of ornithine. In addition, if the cellular extract contains ornithine in significant concentrations, then the calculation of the specific activity of the added ornithine would be difficult. Moreover, it is difficult to get high specific activity [^{14}C]-ornithine and ^{14}C labeling is relatively expensive.

1.3.2.2. Determination of Putrescine

Radioactive ornithine can serve as a substrate for ODC and its activity can be determined by the chromatographic and electrophoretic methods described in **Subheading 1.3.1.**. Djurhuus *(14)* used phosphocellulose paper strips as an ion-exchange matrix for binding putrescine, formed during the decarboxylation reaction. Putrescine, which is positively charged, is tightly bound to the phosphocellulose paper, unlike ornithine, which is washed off by rinsing the paper with 0.1*M* ammonia. Tritium-labeled ornithine (which can be purchased at high specific activity at a relatively low cost and is less dangerous than ^{14}C-labeled compound) can be used for this reaction. It should, however, be remembered that putrescine formed during the reaction could be further oxidized by a reaction catalyzed by diamine oxidase. To block this reaction, the inhibitor aminoguanidine could be added to the reaction mixture. This method, which is described here, is very simple and suitable for routine assays.

2. Materials

1. [^3H]Ornithine (American Radiolabeled Chemicals, St. Louis, MO), 55 Ci/mmol; 1 mCi/mL.
2. Whatman P81 paper (Whatman, Maidstone, UK).
3. Scintillation fluid (Zinsser Analytic, Germany).
4. Ammonia (BDH, UK): 0.1*M*.
5. Phosphate-buffered saline (PBS): NaCl (analytical, Frutarom, Haifa, Israel) 85 g/L (1.45*M*), $NaH_2PO_4 \cdot H_2O$ (Merck, Darmstadt, Germany) 18.1 g/L (13 m*M*), $Na_2HPO_4 \cdot 7H_2O$ (Merck) 124 g/L (46 m*M*).
6. ODC buffer: 50 m*M* pyridoxal phosphate (Sigma), 0.1 m*M* EDTA (BDH), 50 m*M* Tris-HCl pH 7.2 (Ultra pure) (Sigma), 5 m*M* dithiothreitol (DTT) (Sigma).
7. 200 m*M* Ornithine (Sigma).

3. Methods

3.1. Assay

ODC activity can be determined in extracts of cells grown in suspension or in monolayers as well as in extracts of plants or tissues. All procedures are done on ice, unless stated otherwise.

1. Cells (prokaryotic, eukaryotic, or plant) grown in suspension:
   a. Sediment by centrifugation.
   b. Wash twice with PBS.
   c. Remove excess PBS by suction.
   d. Resuspend the sedimented cells in 400 μL of ODC buffer.
2. Cells grown as monolayers:
   a. Wash the monolayer twice with PBS.
   b. Remove the excess PBS.

   c. Add ODC buffer (400 μL to each 100-mm diameter plate).
   d. Scrape the cells off using a rubber policeman.
   e. Transfer the cells to an Eppendorf tube.
   f. Disrupt the cells by freezing (at –70°C for at least 20 min) and thawing (at 37°C for times as short as possible; *see* **Note 1**).
   g. Repeat this procedure four times.
3. Plant or animal tissues:
   a. Wash tissues in PBS.
   b. Suspend in ODC buffer 1:5 (w/v).
   c. Extracted the enzyme from the tissues by four cycles of freezing and thawing.
4. Centrifuge at 4°C for 20 min at 15,000*g* to sediment cellular debris.
5. Remove the supernatant fluid containing ODC and transfer to a new tube.
6. Set aside an aliquot of each sample for protein assay ***(15)***.
7. Set up assay tubes as follows:
   a. Supernatant fluid, or ODC buffer in the blank: 100 μL.
   b. Ornithine (200 m*M*): 10 μL.
   c. [^{3}H]ornithine: 10 μL.
8. Incubate at 37°C for 1 h.
9. Stop the reaction by placing the tubes in an ice bath.
10. Spot the reaction mixtures onto phosphocellulose P81 papers (5.5 × 1.5 cm for each sample).
11. Spot 10 μL of [^{3}H]ornithine on P81 paper as a control.
12. Attach the papers to a holder (e.g., a picture frame).
13. Dry the samples onto the paper (in air or oven).
14. Place the frame holding the papers in a large beaker (*see* **Note 2**).
15. Wash for 45 min in 5 L of 0.1*M* ammonia using a magnetic stirrer (*see* **Note 3**)
16. Replace the wash solution with a further 5 L of 0.1*M* ammonia and wash for another 30 min.
17. Wash the papers in distilled water for 30 min.
18. Dry the papers in a vacuum oven at 80°C.
19. Place each paper in a scintillation vial containing scintillation fluid.
20. Count the radioactivity in a scintillation counter (*see* **Notes 4–6**).

3.2. Calculation of Results

ODC activity is expressed as pmol/mg protein. For example, if the specific activity of [^{3}H]ornithine used was 55 Ci/mmol, then 55 mCi ≡ 1000 pmol; 1 μCi is used for each reaction = 1000/55 = 18 pmol; 10 μL of 200 m*M* cold ornithine is used for each reaction (2000 pmol; then [R], the ratio between cold and radioactive ornithine = 2000/18 = 111.11. Hence:

$$(\text{cpm sample} - \text{cpm blank}) \times 18\ \text{pmol}/(\text{cpm standard} - \text{cpm blank}) \times [R]\ (111.11) \times 1/(\text{protein})^* = \text{pmol/mg protein}$$

*The protein is calculated as mg/sample (=mg/100 μL lysate).

4. Notes

1. Because of the lability of the enzyme the lysate should not be kept at 37°C (during the thawing) for longer than necessary.
2. To avoid tearing of the papers by the magnetic stirrer, the holder should be taped to the 5 L beaker.
3. The washing has to be done in a hood because of the evaporation of the ammonia.
4. If high readings are obtained in the controls ([^{3}H]ornithine), papers should be washed again in the ammonia solution.
5. If high results (in all the samples) are obtained (high counts relative to the standard), then the following steps should be considered:
   a. Dilute the sample (lysate) before adding it to reaction mixture (more than 400 μL ODC buffer should be added to the cells); or
   b. Add more cold (nonradioactive) ornithine. If radioactivity is still high after doing one of these steps, then both should be attempted.
6. If low results occur, less cold ornithine or less ODC buffer should be used.

References

1. Blackshear, P. J., Manzella, J. M., Stumpo, D. J., Wen, L., Huang, J. K., Oyen, O., and Young, W. S. (1989) High level, cell-specific expression of ornithine decarboxylase transcripts in rat genitourinary tissues. *Mol. Endocrinol.* **3,** 68–78.
2. Laemmli, U. K. (1970) Cleavage of structural proteins during the assembly of the head of bacteriophage T4. *Nature* **227,** 680–685.
3. Tipnis, U. R., Steiner, A. L., Skiera, C., and Haddox, M. K. (1989) Immunolocalization of ornithine decarboxylase in rat atria. *J. Mol. Cell. Cardiol.* **21,** 743–750.
4. Anehus, S., Emanuelsson, H., Persson, L., Sundler, F., Scheffler, I. E., Heby, O. (1984) Localization of ornithine decarboxylase in mutant CHO cells that overproduce the enzyme. Differences between the intracellular distribution of monospecific ornithine decarboxylase antibodies and radiolabeled alpha-difluoromethylornithine. *Eur. J. Cell. Biol.* **35,** 264–272.
5. Shayovits, A. and Bachrach, U. (1994) Immunohistochemical detection of ornithine decarboxylase in individual cells: potential application for *in vitro* chemosensitivity assays. *J. Histochem. Cytochem.* **42,** 607–611.
6. Gale, E. F. (1940) Enzymes concerned in the primary utilization of amino acids by bacteria. *Bact. Rev* **4,** 135–176.
7. Tabor, C. W., Tabor, H., and Bachrach, U. (1964) Identification of the aminoaldehydes produced by the oxidation of spermine and spermidine with purified plasma amine oxidase. *J. Biol. Chem.* **239,** 2194–2203.
8. Seiler, N. and Wiechman, M. (1965) Zum Nachweis von Aminen in 10-10-MolMasstab: Trennung von 1-Dimethylam Trennung von 1-Dimethyl-aminohaphthalin-5-sulfonsaunreamiden auf Dunnschichtchro-matogrammen. *Experientia* **21,** 203,204.
9. Fischer, F. G. and Bohn, H. (1957) Uber die Bestimmung von Spermin, Spermidin und anderen biogenen Aminen nach papiere-lektrophoretischer Abtrennung und ihre Mengen-verhaltnisse in tierischen Organen *Hoppe Seylers. Z. Physiol. Chem.* **308,** 108–115.

10. Rosenthal, S. M. and Tabor, C. W. (1956) The pharmacology of spermine and spermidine: distribution and excretion. *J. Pharmacol. Exp. Ther.* **116,** 131–138.
11. Marton, L. J. and Lee, P. L. Y. (1975) More sensitive automated detection of polyamines in physiological fluids and tissue extracts with o-pthalaldehyde. *Clin. Chem.* **21,** 1721–1724.
12. Marton, L. J., Heby, O., Wilson, C. B., and Lee, P. L. Y. (1974) An automated micromethod for quantitative analysis of di- and polyamines utilizing a sensitive high pressure liquid chromatographic procedure. *FEBS Lett* **41,** 99–103.
13. Russell, D. and Snyder, S. H. (1968) Amine synthesis in rapidly growing tissues: Ornithine decarboxylase activity in regenerating rat liver, chick embryo and various tumors. *Proc. Natl. Acad. Sci. USA* **60,** 1420–1427.
14. Djurhuus, R. (1981) Ornithine decarboxylase (EC 4.1.1.17) assay based upon the retention of putrescine by a strong cation-exchange paper. *Anal. Biochem* **113,** 352–355.
15. Bradford, M. M. (1976) A rapid and sensitive method for the quantitation of microgram quantities of protein utilizing the principle of protein-dye binding. *Anal. Biochem.* **72,** 248–254.

3

Assay of Mammalian Ornithine Decarboxylase Activity Using [^{14}C]Ornithine

Catherine S. Coleman and Anthony E. Pegg

1. Introduction

L-Ornithine decarboxylase (EC 4.1.1.17) (ODC) catalyzes the conversion of L-ornithine to putrescine and CO_2. ODC is dependent on pyridoxal 5'-phosphate (PLP) and thiol-reducing agents for activity *(1)*. At least two key active site residues of ODC are known. Lysine 69 was recently identified as the residue in mouse ODC that forms a Schiff base with PLP and α-difluoromethylornithine, an enzyme-activated irreversible inhibitor of ODC, forms adducts at both lysine 69 and cysteine 360 *(2)*. It has been shown that mutation of cysteine 360 to alanine (C360A) results in an enzyme not only having greatly reduced activity, but the activity of this mutant is much more stable than wild-type ODC in the absence of dithiothreitol. The high dependence of ODC on thiol-reducing agents for activity is, therefore, likely to be a result of a need to maintain cysteine 360 in a reduced state *(3)*.

The assay procedure described in this chapter is used to determine mammalian ODC activity. It relies on the measurement of $^{14}CO_2$ evolved from the enzymatic decarboxylation of L-[1-^{14}C]-ornithine in a reaction mixture that contains pyridoxal 5'-phosphate as the cofactor and the reducing agent dithiothreitol. The $^{14}CO_2$ released is absorbed by hyamine hydroxide contained in a center well suspended from a rubber stopper that seals the reaction test tube. The center well is removed at the end of the incubation period and transferred into a toluene-based scintillation fluid for counting by liquid scintillation spectrometry.

From: *Methods in Molecular Biology, Vol. 79: Polyamine Protocols*
Edited by: D. Morgan Humana Press Inc., Totowa, NJ

2. Materials

2.1. Equipment

1. Glass assay tubes, 7 mL, 75 × 16 mm internal diameter (Beckton Dickinson, Rutherford, NJ).
2. Rubber stoppers (Kontes Scientific Glassware, Vineland, NJ).
3. Polypropylene center wells (Kontes Scientific Glassware).
4. Shaking water bath at 37°C.
5. Glass scintillation vials.
6. Toluene-based scintillation fluid.

Note: Gloves should be worn when handling scintillation fluid and all manipulations should be done in a fume hood.

2.2. Reagents

All aqueous solutions should be prepared using distilled, deionized water.

1. Enzyme extract in suitable buffer e.g., purified ODC is typically prepared in Buffer A: 25 m*M* Tris-HCl, pH 7.5, containing 2.5 m*M* dithiothreitol, 0.1 m*M* EDTA, and 0.02% (v/v) Brij 35 solution. Cell extracts can be prepared in Buffer B: 10 m*M* Tris-HCl, pH 7.5, 2.5 m*M* dithiothreitol, and 0.1 m*M* EDTA. **Note:** Freshly prepared buffer solutions containing dithiothreitol can be stored at 4°C and are stable for up to 5 d at this temperature.
2. 1*M* Tris-HCl buffer, pH 7.5, at 37°C: Use ultrapure grade Tris base. Adjust to the appropriate pH at 37°C using concentrated HCl.
3. 250 m*M* dithiothreitol: This solution is unstable at room temperature but can be stored in suitable aliquots at –20°C.
4. 2 m*M* pyridoxal 5'-phosphate: store in suitable aliquots at –20°C in foil-covered tubes. Note that pyridoxal 5'-phosphate should be prepared in dim light and the solution stored in a foil-covered container to prevent photodecomposition.
5. 20 m*M* L-ornithine: Store in suitable aliquots at –20°C
6. L-[1-^{14}C] ornithine (0.1 mCi/mL ; specific activity ~50 mCi/mmol, Dupont NEN, Boston, MA). As with all radioactive materials, gloves should be worn and containment precautions taken when handling. Store at –20°C.
7. 1*M* hyamine hydroxide in methanol (scintillation grade, Research Products International Corp., Mount Prospect, IL). Store at 4°C.
8. 5*M* sulfuric acid—store at room temperature.

3. Methods

All enzyme extracts should be prepared at 4°C. Reaction tubes and all assay solutions should be chilled and kept on ice prior to the incubation.

1. Prepare a stock assay mixture in which each 50 µL aliquot to be added to one reaction tube consists of: 12.5 µL of 1*M* Tris-HCl, pH 7.5, 5 µL of 2 m*M* pyri-

doxal 5'-phosphate, 2.5 µL of 250 m*M* dithiothreitol, 5 µL of 20 m*M* L-ornithine, and 2.5 µL of L-[1-^{14}C] ornithine, and 22.5 µL of deionized water.

2. The ODC assay is done in a total volume of 250 µL. To each prechilled tube add an appropriate volume of deionized water and/or sample buffer and a 50 µL aliquot of assay mix as described in **step 1**. Add the enzyme extract last, making up the desired 250 µL final volume. Duplicate blank tubes should be included in which water or buffer is substituted for enzyme extract.
3. Immediately close each reaction tube with a rubber stopper carrying a polypropylene well containing 0.25 mL of 1*M* hyamine hydroxide. Take care to place the stopper firmly into each tube to avoid a tendency to pop out during the course of the reaction.
4. Incubate the reactions at 37°C for 30 min in a shaking water bath.
5. After the 30 min incubation, place the reaction tubes on melting ice. Use a 1-mL syringe to inject 0.3 mL of 5*M* sulfuric acid through the rubber cap (*see* **Note 1**). This allows the $^{14}CO_2$ to be released from the assay medium during a further 30 min incubation at 37°C. The tubes should be left to incubate for at least 30 min to ensure complete absorption of the $^{14}CO_2$ by the hyamine hydroxide.
6. After all the $^{14}CO_2$ has been released, remove the well from the reaction tube and wipe any condensation that has formed on the outside with a tissue. Transfer the well to a glass scintillation vial using scissors to clip the well-stem free of the rubber stopper (*see* **Note 2**).
7. Working in a fume hood, add 10 mL of a toluene-based scintillation fluid and count the samples in a liquid scintillation counter. Include duplicate vials that contain 0.25 mL of hyamine hydroxide, 10 mL of the scintillation fluid, and a 5-µL aliquot of the assay mix described in **step 1** to determine a specific activity for ornithine used in the assay (*see* **Note 3**).

4. Notes

1. One of the most common problems with this assay occurs when adding the sulfuric acid to release the $^{14}CO_2$ from the reaction. To avoid acid contamination of the hyamine, hold the reaction tube firmly and direct the syringe needle through the rubber stopper away from the center well. Also take care to avoid spilling hyamine from the center well into the reaction mixture. A spoiled reaction is obvious by the opaque appearance of the reaction.
2. Radioactive waste disposal: Neutralize the pooled reaction mixtures containing [1-^{14}C]-ornithine and sulfuric acid with solid sodium bicarbonate in a beaker prior to disposal in a suitable liquid waste container.
3. Definition of unit: One unit is defined as the amount of $^{14}CO_2$ produced in µmol/min at 37°C. Specific activity is expressed as µmol/min/mg of protein.

References

1. Pegg, A. E. and Williams-Ashman, H. G. (1981) Biosynthesis of putrescine, in *Polyamines in Biology and Medicine* (Morris, D. R. and Marton, L. J., eds.), Marcel Dekker Inc., New York, pp. 3–42.

2. Poulin, R., Lu, L., Ackermann, B., Bey, P., and Pegg, A. E. (1992) Mechanism of the irreversible inactivation of mouse ornithine decarboxylase by α-difluoromethylornithine. *J. Biol. Chem.* **267,** 150–158.
3. Coleman, C. S., Stanley, B. A., and Pegg, A. E. (1993) Effect of mutations at active site residues on the activity of ornithine decarboxylase and its inhibition by active site-directed irreversible inhibitors. *J. Biol. Chem.* **268,** 24,572–24,579.

4

Assay of Mammalian S-Adenosylmethionine Decarboxylase Activity

Lisa M. Shantz and Anthony E. Pegg

1. Introduction

S-adenosylmethionine decarboxylase (AdoMetDC; EC 4.1.1.50) catalyzes the conversion of *S*-adenosylmethionine to *S*-5'-deoxyadenosyl-(5')-3-methylthiopropylamine (decarboxylated S-adenosylmethionine) and CO_2. The enzyme is unusual among decarboxylases in that it does not use pyridoxal phosphate as a cofactor, but rather uses a covalently bound pyruvate formed from an internal serine residue *(1)*. The mammalian enzyme is synthesized as a 38 kDa proenzyme, which is cleaved in an apparently autocatalytic processing reaction to form the 7.7 and 30.6 kDa subunits of the mature enzyme, with the pyruvate cofactor covalently attached to the N-terminus of the larger subunit *(2)*. The small subunit remains in the native form of the enzyme, which appears to exist as an $\alpha_2\beta_2$ heterotetramer *(3)*. In addition, the small subunit has been shown to contain residues essential for catalytic activity *(4)*.

Although the *Escherichia coli* AdoMetDC, which shows very little homology to the mammalian enzyme, is Mg^{2+}-dependent, both the activity and the proenzyme processing of the mammalian enzyme have been shown to be strongly activated by putrescine *(5,6)*. It is also known that mammalian AdoMetDC requires thiols for maintenance of maximal activity, probably because of an essential cysteine residue at the active site *(7)*. The assay procedure described in this chapter has, therefore, been optimized for the mammalian enzyme. AdoMetDC activity is most easily assayed by measuring $^{14}CO_2$ release from *S*-adenosyl-L-[carboxy-^{14}C] methionine. The evolved

From: *Methods in Molecular Biology, Vol. 79: Polyamine Protocols*
Edited by: D. Morgan Humana Press Inc., Totowa, NJ

$^{14}CO_2$ is absorbed by a solution of hyamine hydroxide and counted in a scintillation counter.

2. Materials

2.1. Equipment

1. Glass assay tubes, 75 × 16 mm internal diameter, approx 7 mL (Becton Dickinson, Rutherford, NJ).
2. Rubber septa (Kontes Scientific Glassware, Vineland, NJ).
3. Polypropylene center wells (Kontes Scientific Glassware).
4. Shaking water bath, 37°C.
5. Toluene-based liquid scintillation counting fluid.
6. Glass scintillation vials.

2.2. Reagent Stock Solutions (Use Distilled, Deionized Water for Preparation of All Solutions)

1. Sodium phosphate buffer, 0.5*M*, pH 7.5. This solution is prepared by combining 16 mL of 0.5*M* monobasic sodium phosphate and 84 mL of 0.5*M* dibasic sodium phosphate diluted to a total volume of 200 mL with water. Store at –20°C. **Note:** Tris inhibits AdoMetDC activity slightly and, therefore, phosphate is the preferred buffer for this assay.
2. Dithiothreitol, 25 m*M*. Store in aliquots at –20°C. The solution is unstable at room temperature and is stable for approx 5 d at 4°C.
3. Putrescine dihydrochloride, 15 m*M*. Store at –20°C.
4. *S*-adenosyl-L-methionine chloride, 4 m*M*. This material is typically about 70% pure when purchased from commercial sources. Therefore, a 4 m*M* solution contains approx 2.8 m*M* *S*-adenosylmethionine, and this number is used in the calculation of the final *S*-adenosylmethionine concentration. This stock solution is extremely unstable at room temperature, with as much as a 10% purity loss per day. Therefore, the solution must be stored at –20°C.
5. *S*-adenosyl-L-[carboxy-^{14}C]methionine, 20 µCi/mL (specific activity approx 56 mCi/mL). Store at –20°C. At present, the only commercial source for this reagent is Amersham Life Science, Inc. (Cleveland, OH).
6. Enzyme extract in suitable buffer. Purified AdoMetDC is typically prepared in 25 m*M* Tris-HCl, pH 7.5, 2.5 m*M* dithiothreitol; 2.5 m*M* putrescine, 0.1 m*M* EDTA, and 0.02% (v/v) Brij 35 solution. Cell extracts can be prepared in 10 m*M* Tris-HCl, pH 7.5, 2.5 m*M* dithiothreitol, and 0.1 m*M* EDTA. **Note:** These buffer solutions contain dithiothreitol and should be stored as described in **item 2**.
7. Sulfuric acid, 5*M*. Store at room temperature.
8. Hyamine hydroxide, 1*M* in methanol. Store at 4°C.

3. Methods

All solutions should be kept at 4°C until the start of the assay. Always wear gloves when working with radioactivity.

1. Make the following reaction mix in a polypropylene tube (e.g., Falcon 2059 tube):

Stock solution	Volume added per assay
0.5*M* Phosphate buffer	25 µL
25 m*M* Dithiothreitol	12.5 µL
15 m*M* Putrescine	50 µL
4 m*M* *S*-adenosylmethionine	12.5 µL
S-adenosyl-L-[carboxy-^{14}C]methionine	5 µL (0.1–0.2 µCi)
Double-distilled water	45 µL

2. Set up assay tubes in a test tube rack on melting ice. Add 150 µL of the above reaction mix to each tube.
3. The reaction is started by the addition of enzyme. Add 100 µL of enzyme solution to each tube. The final reaction volume of all tubes should be 250 µL (*see* **Note 1**). Also include duplicate blank reactions that substitute 100 µL of water or buffer for enzyme solution.
4. Immediately cap tubes with rubber septa that carry a polypropylene center well containing 250 µL of 1*M* hyamine hydroxide (*see* **Note 2**).
5. Incubate tubes in a 37°C water bath, shaking gently (approx 50 rpm) for 30–60 min (*see* **Note 3**).
6. Stop the reaction by placing the assay tubes on ice. Using a 1 mL syringe, inject 300 µL of 5*M* sulfuric acid through the rubber septum and into the reaction mix. Acidification of the reaction mix releases $^{14}CO_2$, allowing it to be trapped by the hyamine solution (*see* **Note 4**).
7. In order to ensure complete absorption of the $^{14}CO_2$, incubate for a further 30 min at 37°C (*see* **Note 5**).
8. Remove the wells containing the hyamine hydroxide solution from each assay tube by clipping them off with a pair of scissors into glass scintillation vials (*see* **Note 6**).
9. Add 10 mL of toluene-based liquid scintillation counting fluid to each vial. Also count 10 µL of the reaction mix from step 1 combined with 250 µL of hyamine hydroxide solution and 10 mL of scintillation counting fluid.
10. Count all samples for 5 min in a scintillation counter using a channel appropriate for ^{14}C (*see* **Note 7**).

4. Notes

1. The volume of enzyme solution added to the reaction mix will vary greatly depending on the purity of the enzyme preparation. Typically for crude cell cytosols 100 µL of lysate is added. In contrast, very small volumes of pure enzyme must be used and should be diluted to 100 µL before adding to the reaction mix. The buffers most commonly used are described in **Subheading 2.2., step 6**. In order to insure a linear rate of reaction, the amount of enzyme added to the assay should be sufficient to convert not more than 10% of the ^{14}C-S-adenosyl methionine to $^{14}CO_2$.

2. Although the reaction should not begin while tubes are on ice, it is best to add enzyme to two or three tubes at a time and immediately cap the tubes. Push the septa firmly into the tubes because the temperature changes within the tubes when they are moved to 37°C can cause the septa to pop out if they are not properly placed.
3. The reaction can be continued for up to 120 min for samples containing very little activity, however 30 min is usually sufficient to obtain counts significantly above background (background counts from blanks average about 30 cpm).
4. A common error in this assay is splashing acid into the hyamine solution. If the hyamine solution comes into contact with the sulfuric acid, it will not absorb the CO_2 from the assay mix and counts will be very low. Therefore, in order to avoid the center well, use of a long needle (e.g., 23 gage 1½ in.) is recommended when injecting acid. Likewise, care must be taken to avoid splashing hyamine from the center wells into the reaction mix. If this occurs, the reaction mix usually turns cloudy and again counts will be very low and must be discarded.
5. This step can be carried out for longer than 30 min, however, absorption of the $^{14}CO_2$ by the hyamine solution is usually complete in 30 min and, therefore, longer incubations do not result in higher counts.
6. Acidification of the reaction mix sometimes causes condensation to form on the center wells during the subsequent 30-min incubation. Therefore, it is recommended that the outside of the center wells be wiped gently before clipping them into the scintillation vials. Because these samples are counted in toluene-based counting fluid, even a small amount of aqueous solution can lower the counting efficiency and result in less reproducible results. Always add counting fluid in a fume hood.
7. Because chemiluminescence is sometimes a problem in counting these samples, samples should either be equilibrated in the dark before counting or counted more than once to ensure accurate and reproducible results.
8. Definition of a unit: One unit of activity of AdoMetDC is defined as the amount of enzyme releasing 1 nmol of $^{14}CO_2$/min at 37°C. Specific activity is expressed as units per milligram of protein.

References

1. Pegg, A. E. (1986) Recent advances in the biochemistry of polyamines in eukaryotes. *Biochem. J.* **234,** 246–262.
2. Shirahata, A., Christman, K., and Pegg, A. E. (1985) Quantitation of S-adenosylmethionine decarboxylase protein. *Biochemistry* **24,** 4417–4423.
3. Pajunen, A., Crozat, A., Jänne, O. A., Ihalainen, R., Laitinen, P. H., Stanley, B. A., Madhubala, R., and Pegg, A. E. (1988) Structure and regulation of mammalian S-adenosylmethionine decarboxylase. *J. Biol. Chem.* **263,** 17,040–17,049.
4. Stanley, B. A. and Pegg, A. E. (1991) Amino acid residues necessary for putrescine stimulation of human S-adenosylmethionine decarboxylase proenzyme processing and catalytic activity. *J. Biol. Chem.* **266,** 18,502–18,506.

5. Kameji, T. and Pegg, A. E. (1987) Effect of putrescine on the synthesis of S-adenosylmethionine decarboxylase. *Biochem. J.* **243,** 285–288.
6. Pegg, A. E. and Williams-Ashman, H. G. (1968) Stimulation of the decarboxylation of S-adenosylmethionine by putrescine in mammalian tissues. *Biochem. Biophys. Res. Commun.* **30,** 76–82.
7. Williams-Ashman, H. G., Jänne, J., Coppoc, G. L., Geroch, M. E., and Schenone, A. (1972) New aspects of polyamine biosynthesis in eukaryotic organisms. *Adv. Enzyme Regul.* **10,** 225–245.

5

Assay of Spermidine and Spermine Synthases

Laurie Wiest and Anthony E. Pegg

1. Introduction

Spermidine and spermine syntheses carry out the transfer of aminopropyl groups from decarboxylated *S*-adenosylmethionine (dcAdoMet) to putrescine and spermidine, respectively. Despite the close similarity of these reactions, they are totally distinct enzymes. Spermidine synthase (E. C. 2. 5.1.16), which is present in virtually all organisms, transfers the aminopropyl group from dcAdoMet to putrescine forming spermidine. Spermine synthase (E. C. 2.5.1.22), which has a much more limited distribution and is absent from many prokaryotes, transfers the aminopropyl group to spermidine forming spermine. The cDNAs for both enzymes have been cloned and expressed and the derived amino acid sequences show little similarity except in the putative dcAdoMet binding site ***(1–3)***.

Both reactions convert dcAdoMet to 5'-methylthioadenosine (MTA) and the measurement of the MTA formed in the presence of the appropriate amine acceptor is the basis of the assay ***(4)***. The dcAdoMet is prepared labeled with either [^{35}S] or [^{3}H] in the 5'-methyl group. After donation of the aminopropyl group, the radioactivity is converted to labeled MTA. MTA, which lacks the strong positive charge contributed by the sulfonium center of dcAdoMet, is readily separated from it by use of a small phosphocellulose column ***(5)*** that retains the dcAdoMet. The eluate is then counted in a scintillation counter. Since the two aminopropyltransferase enzymes are specific for the appropriate amine acceptor, supplying the specific substrate allows assaying the aminopropyltransferase of interest using the same procedure.

The assay itself is, thus, very simple but the procedure is complicated by the lack of commercial sources for dcAdoMet which must, therefore, be prepared in the laboratory. The [^{35}S]dcAdoMet has the advantage of a very high specific

From: *Methods in Molecular Biology, Vol. 79: Polyamine Protocols*
Edited by: D. Morgan Humana Press Inc., Totowa, NJ

activity allowing assays of very small samples and of price, but requires two enzymes for its preparation [*S*-adenosylmethionine synthetase (MAT) and *S*-adenosylmethionine decarboxylase (AdoMet DC)]. Since [^{3}H-*methyl*]AdoMet is commercially available, [^{3}H-*methyl*]dcAdoMet can be prepared with only the AdoMetDC, but this is a much more expensive option. The cDNAs for both MAT and AdoMetDC have been cloned and expressed in *Escherichia coli* using vectors that give a very high yield of active protein. Purification of the enzymes from these sources may be desirable, but is not essential for a good yield of product. The Mg^{2+}-dependent AdoMetDC from *E. coli* is preferable to the mammalian enzyme since it is less product inhibited and thus gives a higher yield of dcAdoMet. This product is readily separated from methionine, MTA, and AdoMet itself by chromatography on a Dowex ion exchange resin and ion exchange high performance liquid chromatography (HPLC).

2. Materials

1. The majority of reagents used are common laboratory reagents and may be purchased from a preferred supplier. Preparation and use of stocks is dictated by standard laboratory practice. Acetonitrile for use in the mobile phase of the HPLC should be HPLC grade.
2. [^{35}S] methionine, specific activity >1000 Ci/mmol, is available from Amersham Life Science (Cleveland, OH), or DuPont NEN (Boston, MA). [^{3}H-*methyl*]AdoMet is available from both these companies at a specific activity of 80 Ci/mmol.
3. AdoMet was purchased from Sigma (St. Louis, MO).
4. Dowex ion exchange resin, Dowex 50-H^+, is available from Sigma. The solid powder is stable for several years at room temperature, as is the suspended slurry when stored in 0.5*N* HCl at 4°C.
5. Disposable plastic columns are available from Kontes Glass Co. (Vineland, NJ).
6. HPLC was performed on a Whatman Partisil 10 SCX column, 4.6 mm × 25 cm, with a 4.6 mm × 5 cm precolumn packed with the same material, packed by the user, but available commercially from the manufacturer. The lifespan of the column is several years if used exclusively for the separation of dcAdoMet from this reaction, but will vary with usage. Heavy use will shorten the lifespan.
7. Phosphocellulose cation exchanger, Cellex-P, is available from Bio-Rad Labs. It is stable as a solid for several years at room temperature, and stored as a slurry in 25 m*M* HCl is stable for several years as well when stored at 4°C.
8. MAT enzyme was obtained from G. D. Markham, Fox Chase Cancer Center, 7701 Burholme Avenue, Philadelphia, PA 19111, USA. This enzyme was prepared using *E. coli* strain DM22pK8, which is available from Dr. Markham ***(6)***.
9. AdoMetDC was obtained from *E. coli* strain HT 383/pSPD1 ***(7,8)***. This strain was provided by H. Tabor, Laboratory of Biochemical Pharmacology, NIADDK, NIH, Bethesda, MD 20205, and it or related strains that give even higher yields can be obtained from him or from our laboratory. Recombinant human AdoMetDC,

which can be obtained using plasmid pINSAM-3 *(9)* available from our laboratory, can be used, but is not recommended (*see* **Note 1**).

3. Methods

3.1. Preparation of Labeled [^{35}S]dcAdoMet

1. The synthesis reaction uses a final concentration of 50 m*M* sodium phosphate buffer, pH 7.5, 100 m*M* KCl, 20 m*M* $MgCl_2$, 500 n*M* ATP, 530 μCi [^{35}S]methionine, 4 μg purified MAT enzyme (*see* **Note 2**), and 15 U of *E. coli* AdoMetDC (*see* **Notes 1** and **3**; 1 unit is defined as the amount of enzyme required to convert 1 nmol of AdoMet to dcAdoMet per minute). Unlabeled methionine may be added if a lower specific activity is desired. The volume of the reaction is typically 175 μL, and may be adjusted as needed (*see* **Note 4**). Incubate the reaction mix at 37°C for 60 min. Terminate the reaction by adding perchloric acid to a final concentration of 4.2%. Allow the mix to sit on ice for 30 min. Spin at 4°C for 10 min at 16,000*g*. Save the supernatant, which may be frozen at this point for storage.
2. Prepare a Dowex 50-H^+, 8 mm × 5 cm column. Wash the Dowex with 6*M* HCl, then with deionized water to equilibrate. Apply the supernatant. The column is washed first with 0.01*M* HCl and then eluted with 6*M* HCl as described below. Count the radioactivity in 2 μL of each fraction, including the initial application, the dilute HCl wash, and 6*M* HCl wash. The first washes with 0.01*M* HCl use five aliquots of 4 mL each. The eluates from these steps should contain high counts in the first and second fraction and then counts will fall quickly. These fractions contain the unreacted [^{35}S]methionine, which can be reused if desired. Elute the dcAdoMet product with 6*M* HCl using 5 aliquots of 3 mL each. Collect and save. Eluted product usually contains high counts in the first three fractions, but more fractions may be collected if necessary. The fractions of the eluted product with the highest counts should be evaporated to dryness. This can be done simply by blowing a stream of air over the fractions in a fume hood. Dispose of radioactive material appropriately (*see* **Note 5**). The dried material may be stored frozen. This fraction contains the desired dcAdoMet plus some AdoMet and MTA.
3. Prepare the HPLC mobile phase, consisting of 0.15*M* ammonium formate, pH 4.0 (adjust the pH with formic acid), plus 10% acetonitrile, degassed and filtered. Equilibrate the Partisil 10 SCX column in this buffer. Dissolve the dried product in a volume suitable for the loop size available. Using a flow rate of 2 mL/min and a temperature between room temperature and 30°C, verify the separation of AdoMet, MTA, and dcAdoMet standards with a UV detector at 254 nm. Using concentrations of 0.2 m*M* for MTA and AdoMet and 0.6 m*M* for dcAdoMet, aliquots of 50 μL will produce satisfactory peaks using a full scale of 0.2. The elution order is MTA, AdoMet, and dcAdoMet. Retention time may vary with the state of the column, but is about 2 min for MTA, 3.5 min for AdoMet, and 12 min for dcAdoMet when using a new column (*see* **Note 6**).
4. Load and run the dissolved material once it is verified that dcAdoMet will separate from the other two peaks. Unless pure labeled AdoMet is needed for another

project, it is not necessary to get a good separation of AdoMet and MTA since these are discarded or reused. Collect 1 min fractions for about 20 min; dcAdoMet is a late eluting peak and tends to tail. Count 2 µL of each fraction. The counts will rise swiftly, then drop, and rise again when the dcAdoMet peak elutes.

5. Take the dcAdoMet fractions with the highest counts and remove the buffer and acetonitrile as follows: Prepare another Dowex 50-H^+ column as detailed above in **Subheading 3.1., step 2**. Load the sample and collect the eluant. Wash with 0.5*M* HCl, seven aliquots of 5 mL each. Then elute the [^{35}S]dcAdoMet with 6*M* HCl, five aliquots of 5 mL each. Count 2 µL of each fraction. The load and wash fractions should have no counts, and dcAdoMet usually elutes in the first three fractions of 6*N* HCl. Dry with forced air and dissolve in 500 µL of deionized water. Calculate yield by comparing the radioactivity present in the final dcAdoMet preparation with that in the methionine starting material. Store frozen in aliquots at –20°C.

3.2. Preparation of [^{3}H-Methyl]dcAdoMet or Unlabeled dcAdoMet

1. Incubate the *E. coli* AdoMetDC (15 U; *see* **Note 1**) with 0.2*M* Tris-HCl, pH 7.5, 0.1*M* $MgCl_2$, 5 m*M* DTT, 0.5 mCi of [^{3}H-*methyl*]dcAdoMet, or 1 µmol AdoMet, for 60 min at 37°C. Terminate the reaction with perchloric acid and proceed as in **Subheading 3.1., steps 2–5**.

3.3. Preparation of Cellex-P (Cellulose Phosphate)

1. Suspend 10–20 g of the Cellex-P in 500 mL of 0.25*M* HCl. Incubate the slurry for 30 min at room temperature, with occasional stirring. Do not stir vigorously to avoid breaking up the material. Filter with a vacuum filter apparatus and wash with deionized water.
2. Resuspend the slurry in 500 mL of 0.25*M* NaOH and incubate 10 min at room temperature. Neutralize with HCl. Filter and rinse as in **step 1**.
3. Repeat the suspension in 0.25*M* HCl step. Wash and rinse as above. Place the Cellex-P in 25 m*M* HCl and let it settle, pouring off the fine particles. Repeat this step twice. This will remove small broken up particles that will make the columns run slowly. Resuspend in 25 m*M* HCl. Keep the slurry volume small, but it should be dilute enough so that it can be pipeted.
4. To prepare the minicolumns for use in the spermidine and spermine synthase assays, place a small piece of glass wool in the bottom of a 5 ¾ in (14.5 cm) glass Pasteur pipet. Add about 1 mL of the Cellex-P slurry. The final column size in the pipet should be about 1 cm. Each reaction sample will require one minicolumn. Prepare extra minicolumns. Store the columns in 25 m*M* HCl at 4°C.
5. Prepare the columns for use by washing each with two aliquots of 2 mL 0.5*M* HCl, followed by one aliquot of 2 mL of deionized water. Equilibrate the columns with three aliquots of 2 mL of 25 m*M* HCl.

3.4. Spermidine and Spermine Synthase Assays

1. To prepare samples, resuspend mammalian or bacterial samples in 50 m*M* sodium phosphate buffer, pH 7.2, 0.3 m*M* EDTA, 10 m*M* β-mercaptoethanol at 4°C. In

order to break the cells, mammalian cells may be sonicated as below for bacterial cells (*see* **Note 7**), but freezing and thawing two times is adequate. Whole tissue samples may be homogenized with a polytron. Bacterial cells should be sonicated with care to keep them at 4°C during the process. Sonication in our laboratory is carried out with a Misonix XL 2020 sonicator using the microtip at maximal power for a total process time of 2 min in a pulsed mode of 10 s on and 10 s off. The sample was kept chilled in an ice-water bath at all times. Spin the samples at 12,000*g* for 30 min at 4°C. Remove and save the supernatant. Determine protein levels by method of choice.

2. Prepare a stock reaction mix according to the number of samples to be assayed in which each 50-µL aliquot to be added to one reaction consists of: 40 µL of 0.5*M* sodium phosphate, pH 7.5, 5 µL of 0.2 m*M* unlabeled dcAdoMet, and 5 µL of [^{35}S]dcAdoMet containing 40,000–100,000 cpm (*see* **Note 8**). Aliquot 50 µL to each tube. Small glass tubes are adequate, large enough to hold at least 1.5 mL.
3. Add 10 µL of 10 m*M* putrescine, if assaying spermidine synthase, or 10 µL of 10 m*M* spermidine if assaying spermine synthase. Add the sample (dilutions may be necessary). The final volume is made up to 200 µL with deionized water.
4. Incubate the samples at 37°C for 30 min. Stop the reaction with 1 mL of 25 m*M* HCl. Do not add additional HCl to the zero time-point.
5. Include a blank with no polyamine, which should be subtracted from the samples. Also include a zero time, identical to the sample, but with 1 mL of 25 m*M* HCl added. There should be little difference between the blank and the zero time assays. If there is a difference, it is probably because of the presence of sufficient spermidine or putrescine in the sample to allow for some reaction to occur, and the samples should be dialyzed to remove this polyamine.
6. The Cellex-P minicolumns can be prepared for use during the incubation step, as detailed in **Subheading 3.3., step 5**. After equilibration, suspend the minicolumns over scintillation vials (*see* **Note 9**). Apply sample and collect eluent in the vial. The volume should be about 1.2 mL. Apply to each column 2 mL of 25 m*M* HCl; collect in the same scintillation vial. Remove the vials, add an appropriate scintillation cocktail that will accept aqueous samples, shake, and count on a channel appropriate for ^{35}S.
7. The assay measures the [^{35}S]MTA produced. Results can be expressed as a percentage of control, or units may be calculated using the specific activity of the labeled dcAdoMet.
8. The minicolumns are reusable, and may be cleaned and stored for future use. Add two aliquots of 2 mL each of 0.5*M* HCl. Apply one aliquot of 2 mL deionized water. Apply one aliquot of 2 mL of 25 m*M* HCl. Store in 25 m*M* HCl at 4°C. These columns are stable for at least 2 yr, or until protein buildup interferes with their function. Columns can be cleaned of protein deposits by washing with two aliquots of 2 mL of 0.1*N* NaOH, followed by re-equilibrating with deionized water. Prepare the columns for use as described in **Subheading 3.3., steps 4** and **5**.

4. Notes

1. Mammalian AdoMetDC may also be used in **Subheading 3.3., step 1**, or **Subheading 3.2., step 1**. instead of the *E. coli* enzyme for the production of dcAdoMet,

but the yield is lower and it is critical to also add putrescine to 0.2 m*M* since this activates mammalian AdoMetDC. It may be necessary to do some preliminary assays to determine the appropriate amount of enzyme to use in the production.

2. The MAT enzyme appears to be stable for several years, and withstands repeated freezing and thawing. Any problems in the synthesis are generally caused by the more unstable AdoMetDC.
3. Poor yield of [^{35}S]dcAdoMet in **Subheading 3.3., step 1** generally results from an insufficient amount of AdoMetDC enzyme in the mix. Use enzyme with high activity.
4. All the reagents for the dcAdoMet synthesis in **Subheading 3.3., step 1** must be in solution. If the reaction is kept at room temperature and the enzymes added last, this should eliminate that problem, or the volume may be adjusted.
5. Remember that the large amounts of radioactivity produced should be disposed of properly and carefully.
6. The HPLC purification is important to obtain pure dcAdoMet but this separation is simple even with old columns. The more demanding separation of MTA from AdoMet is not critical unless these products are needed for another procedure.
7. Exercise great care when sonicating mammalian cells. Too much sonication will inactivate the enzyme. The double freeze/thaw procedure is a better alternative.
8. As the ^{35}S decays, the volume of material added to equal 40,000 cpm must be adjusted.
9. When using the minicolumns for the spermidine and spermine synthase assays, a square of Plexiglas drilled with holes, but leaving a shoulder for the column to sit on, is very convenient. Holes can be spaced over the scintillation vials, allowing each column to drip into a vial.

References

1. Tabor, C. W. and Tabor, H. (1987) The speEspeD operon of *Escherichia coli.* Formation and processing of a proenzyme form of S-adenosylmethionine decarboxylase. *J. Biol. Chem.* **262,** 16,037–16,040.
2. Myöhänen, S., Wahlfors, M., Alhonen, L., and Jänne, J. (1994) Nucleotide sequence of mouse spermidine synthase cDNA. *DNA Sequence* **4,** 343–346.
3. Korhonen, V. , Halmekytö, M., Kauppinen, L., Myöhänen, S., Wahlfors, J., Keinänen, T., Hyvönen, T., Alhonen, L., Eloranta, T., and Jänne, J. (1995) Molecular cloning of a cDNA encoding human spermine synthase. *DNA Cell Biol.* **14,** 841–847.
4. Pegg, A. E. (1983) Assay of aminopropyltransferases, in *Methods in Enzymology, vol. 94, Polyamines* (Tabor, H. and Tabor, C. W., eds.), Academic, New York.
5. Raina, A., Eloranta, T., and Pajula, R. (1983) Rapid assays for putrescine aminopropyltransferase (spermidine synthase) and spermidine aminopropyltransferase (spermine synthase), in *Methods in Enzymology, vol. 94, Polyamines* (Tabor, H. and Tabor, C. W., eds.), Academic, New York.
6. Markham, G. D., Parkin, D. W., Mentch, F., and Schramm, V. L. (1987) A kinetic isotope effect study and transition state analysis of the *S*-adenosylmethionine synthetase reaction. *J. Biol. Chem.* **262,** 5609–5615.

7. Markham, G. D., Tabor, C. W., and Tabor, H. (1982) *S*-adenosylmethionine decarboxylase of Escherichia coli. *J. Biol. Chem.* **257,** 12,063–12,068.
8. Shirahata, A., Christman, K., and Pegg, A. E. (1985) Quantitation of *S*-adenosylmethionine decarboxylase protein. *Biochemistry* **24,** 4417–4423.
9. Shantz, L. M., Stanley, B. A., Secrist, J. A., and Pegg, A. E. (1992) Purification of human *S*-adenosylmethionine decarboxylase expressed in *Escherichia coli* and use of this protein to investigate the mechanism of inhibition by the irreversible inhibitors, 5'-deoxy-5'-[(3-hydrazinopropyl)methylamino]adenosine and 5'{[(Z)-4-amino-2-butenyl]methylamino-5'-deoxyadenosine. *Biochemistry* **31,** 6848–6855.

6

Measurement of Spermidine/Spermine N^1-Acetyltransferase Activity

Heather M. Wallace and Deborah M. Evans

1. Introduction

The breakdown of the polyamines, spermidine and spermine (**Fig. 1**), is a two-step process involving in the first reaction, spermidine/spermine N^1-acetyltransferase (abbreviated to N^1-SAT or SSAT), and in the second reaction, polyamine oxidase (PAO). N^1-SAT is the rate-limiting enzyme in this catabolic pathway. It is a highly inducible enzyme whose activity is increased by a range of hormones and drugs *(1–3)*. N^1-SAT acetylates primary amino groups separated from another nitrogen atom by a three-carbon aliphatic chain *(4)*. The reaction appears to occur by an ordered Bi Bi mechanism. The enzyme acetylates both spermine and spermidine, but it is unclear which of these polyamines is the preferred substrate in vivo.

Acetylation of the polyamines reduces the number of positive charges and may facilitate the release of polyamines from negatively charged, nuclear binding sites *(5)*. In the case of DNA, removal of polyamines that act to condense and stabilize the chromatin structure may result in an alteration of function or activity. Less condensed DNA may, for example, be more readily degraded by nucleases present within the cell.

The loss of charge also alters the polarity of the polyamine, which may facilitate the transport of polyamines across the plasma membrane. Indeed the excretion of N^1-acetylspermidine has been shown to be a specific, highly regulated process that results in a decrease in intracellular polyamine content *(6–8)*. Increased polyamine efflux is associated with the inhibition of cell growth by nutrient deprivation *(9)*, density dependent inhibition of growth *(6)*, or cytotoxic drugs *(10)*. This removal of polyamines is probably a cytoprotective mechanism that prevents the concentration of free polyamine reaching cyto-

From: *Methods in Molecular Biology, Vol. 79: Polyamine Protocols*
Edited by: D. Morgan Humana Press Inc., Totowa, NJ

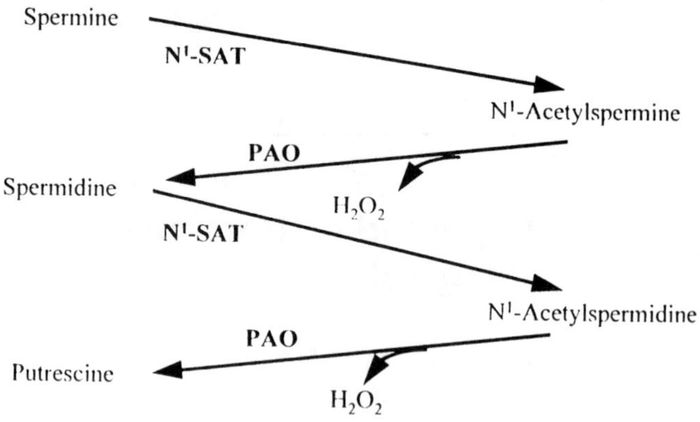

Fig. 1. The polyamine catabolic pathway.

toxic levels ***(11)***. In humans, polyamines are excreted in the urine mainly as monoacetylated derivatives ***(12)***. This urinary output is increased in a number of pathological conditions, such as cancer ***(13,14)*** and psoriasis ***(15)***, and in normal physiological conditions, such as pregnancy ***(14)***, possibly as the result of increased cell turnover.

The acetylase/oxidase pathway **(Fig. 1)** also provides a means of recycling the polyamines back to putrescine which can, in turn, be reused to synthesize spermidine and spermine. The amount of putrescine that is normally regenerated by the cell was quantified by Seiler ***(16)*** by measuring the increase in urinary polyamine output in the presence of inhibitors of PAO and copper-containing amine oxidases. More recent studies have suggested that regeneration of putrescine is a protective mechanism in cells against hypo-osmotic shock ***(17,18)***.

1.1. Principle

The basis of the SAT assay is that the enzyme activity of a given sample can be calculated by measuring the incorporation of radioactivity from radiolabeled acetyl Coenzyme A into monoacetylspermidine, the end product. This can then be quantified because the product is retained on cellulose phosphate paper that can be measured by liquid scintillation spectroscopy. Individual acetyl derivatives can also be separated and quantified by HPLC with radiomatic detection **(Fig. 2)**.

The measurement of SAT activity has been modified by various groups since the original publication by Libby in 1978 ***(19)***. The method described in **Subheading 3.** has been used successfully in our laboratory to measure SAT activity in tissue extracts and from cell lines.

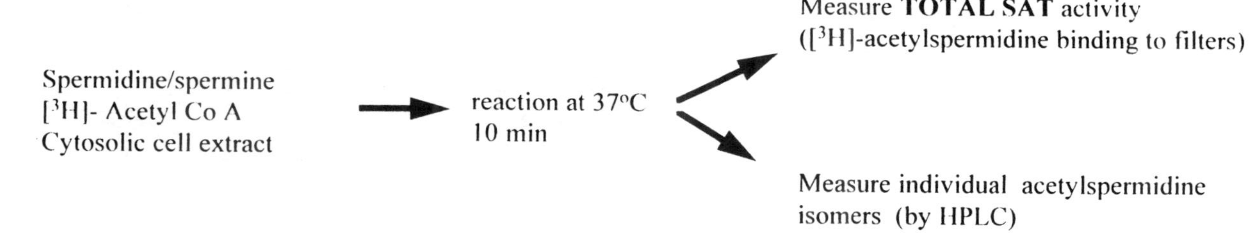

Fig. 2. Principle of the SAT assay.

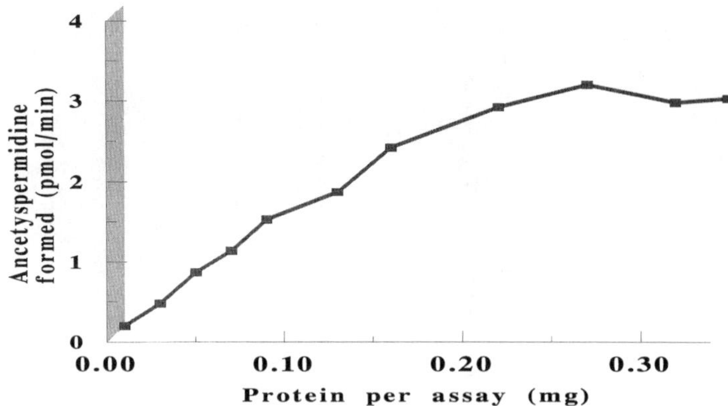

Fig. 3. Relationship between protein content and SAT activity.

1.2. Background

The method was adapted to optimize conditions for measuring polyamine acetylation in human cells. One of the main problems with the method was that it gave low SAT activities and lacked reproducibility. A preliminary study using the hypotonic Tris/HCl homogenizing buffer supplemented with 1 m*M* EDTA and 2.5 m*M* dithiothreitol gave improved results. We chose to use [^{3}H]-labeled acetyl CoA as a cofactor because it was cheaper than [^{14}C]-acetyl CoA. However, in order to maintain an approximately equivalent specific activity the concentration of [^{3}H]-acetyl CoA was increased to 25 μ*M* in the assay. We found that the assay was linear with increasing protein concentration up to 120 μg/assay **(Fig. 3)** and that incubation times longer than 10 min resulted in a loss of linearity.

2. Materials

1. Hypotonic homogenizing buffer: 10 m*M* Tris-HCl (pH 7.5 at 4°C), 2.5 m*M* dithiothreitol, and 1 m*M* EDTA.
2. Phosphate-buffered saline (PBS): Solutions A, B, and C are mixed in the ratio 8:1:1 and stored at 4°C.
3. Solution A: KCl, 0.03% (w/v); NaCl, 1.0% (w/v); Na_2HPO_4, 0.14% (w/v); KH_2PO_4, 0.03% (w/v). The pH is adjusted to 7.2 at 25°C.
4. Solution B: $CaCl_2 \cdot 2H_2O$, 0.1% (w/v).
5. Solution C: $MgCl_2 \cdot 6H_2O$, 0.1% (w/v).

2.1. Equipment

1. Bench-top centrifuge (MSE, Fisons, Crawley, UK).
2. Ultracentrifuge (Centrikon T-1160 centrifuge, Kontron Instruments, Watford, UK).
3. Shaking water bath (standard size).

4. Scintillation counter (Canberra Packard TRI-CARB 1900 CA liquid scintillation spectrometer).
5. Hand held glass homogenizer (Jencons Scientific Instruments Ltd., Leighton Buzzard, UK).
6. Ultra Turrax tissue homogenizer (Janke & Keinkel, Staufen, Germany).
7. Filter trays (7 × 32 cm; with 24 wells of 3 cm diameter).
8. Drying oven (standard).
9. 1.5 mL plastic conical tubes with lids (Eppendorf).
10. Boiling water bath.

2.2. Analysis of Individual Isomers of Spermidine (N^1- and N^8-Acetylspermidine)

1. A reverse-phase μ Bondapack C18 (3.9 × 300 mm 10 μm particle size) HPLC column.
2. Two pumps (Waters Ltd. [Watford, UK] model 501) with an inbuilt manometric module.
3. A Waters Intelligent Sample Processor (WISP 712) autoinjector.
4. An in-line precolumn filter unit.
5. Postcolumn Waters pump (model 6000a solvent delivery system).
6. Fluorescence spectrophotometer (Waters, 420-AC); fluorescence excitation was achieved at 347 nm and emission was measured at 465 nm.
7. Flo-1β radiomatic detector (Canberra Packard, Pangbourne, UK).
8. HPLC buffers were vacuum filtered through 0.45 μm (HA/HV) Millipore filters (Watford, UK) and subsequently degassed with helium prior to use.
   a. Buffer A: 0.1*M* sodium acetate (adjusted to pH 4.5 with glacial acetic acid) containing 10 m*M* 1-octane sulfonic acid.
   b. Buffer B: 0.2*M* sodium acetate (adjusted to pH 4.5 with glacial acetic acid) containing 10 m*M* 1-octane sulfonic acid and acetonitrile (10:3 v/v).
   c. Postcolumn derivatization buffer: OPA reagent/L of distilled water: 50 g boric acid, 44 g potassium hydroxide, 3.5 mL Brij-35 (30% w/v) solution, 2 mL 2-mercaptoethanol, 0.4 g 3-phthaldehyde dissolved in 5 mL of methanol.

3. Methods

3.1. Cell Harvesting

1. Prepare homogenizing buffer on the day of use. Hypotonic buffer consisting of 10 m*M* Tris-HCl (pH 7.5 at 4°C), 2.5 m*M* dithiothreitol, and 1 m*M* EDTA.
2. Ensure PBS and homogenizing buffer are at 4°C
3. Cells required for the acetyltransferase assay are grown on plastic tissue culture plates (10 cm diameter). Remove the medium from the plate and wash the cell sheet twice with ice-cold PBS. Then harvest the cells in 2 × 0.5 mL PBS by gently scraping the cell layer with a "rubber policeman" or cell scraper and transfer the cell suspension to a 1.5 mL Eppendorf tube placed on ice.
4. Sediment the cells at 12,000g_{av} for 2–3 min in a bench-top centrifuge. Discard the supernatant and use the resultant cell pellet to produce a cytosolic fraction.

5. Resuspend the cell pellet in the homogenizing buffer (0.5 mL) and leave the cell suspension on ice for 15 min to allow the cells to swell prior to homogenization. For homogenizing cells a hand-held glass homogenizer (Jencons, Scientific Instruments Ltd., 1 mL) is perfectly adequate, however for homogenizing tissue extracts an Ultra Turrax should be used.
6. A cytosolic extract is prepared by centrifuging the homogenate for 1 h at 100,000g_{av} (325,000 rpm) at 4°C in a Centrikon T-1160 centrifuge. Special high density polypropylene adapters to take 3 × 1 mL polycarbonate centrifuge tubes were used in 12 × 25 mL aluminum rotor.
7. Reserve an aliquot of the cytosolic fraction from each sample to be kept for protein analysis. Samples should be acid extracted in 0.2*M* perchloric acid and then the protein solubilized in 0.3*M* NaOH at 37°C for a minimum of 1 h. The protein content was measured by the method of Lowry et al. ***(20)***.
8. Assay the cytosolic fraction **immediately** for acetyltransferase activity. (If storage of the cytosol is necessary, the stability of the enzyme activity after storage at the available temperature should first be evaluated. In our hands, storage of the cytosol at –70°C resulted in an inconsistent loss of activity and an increase in the variability between results.)

3.2. The Assay Procedure

Assays are always carried out in triplicate to estimate variation within the assay.

1. Place 60 µL of cytosol preparation into an Eppendorf tube and add 10 µL of 30 m*M* spermidine (stock solution), and 10 µL of 1*M* Tris-HCl (stock solution), pH 7.8, at 37°C. The final concentrations in the assay are 3 m*M* spermidine and 100 m*M* Tris buffer (*see* **Note 1**).
2. Place the assay tubes in a water bath at 37°C and initiate the enzyme reaction after 2 min by adding 20 µL of [^{3}H]-acetylCoA mix containing two parts 25 µ*M* cold acetyl CoA and one part [^{3}H]-acetyl CoA on a 10-s cycle.
3. Allow the reaction to proceed for exactly 10 min, then terminate by adding 20 µL of 1*M* hydroxylamine hydrochloride, again on a 10-s cycle, and placing the reaction tube directly on ice. Boil the samples for 3 min to precipitate any protein that can be removed by centrifugation at 12,000g_{av} for 2 min.
4. Spot duplicate 30 µL aliquots of the assay products on to 2.6 cm diameter phosphate cellulose filters (P81) and allow to dry in a hot air oven. Then wash the filters once in tap water, three times with distilled water for 2 min each, and finally once in 100% ethanol to remove unbound [^{3}H]-acetyl CoA. Finally, dry the filters in a hot air oven.
5. The positively charged reaction product, [^{3}H]-acetylspermidine, is retained by the negatively charged paper filters and quantified by counting the filters in a liquid scintillation spectrometer in vials containing 4 mL of Opti-Fluor O scintillation fluid that is suitable for nonaqueous scintillation counting (*see* **Fig. 2**).
6. The activity of the enzyme per assay is expressed as pmol of total acetylspermidine formed/min/mg of protein, taking into account the volume assayed and the original homogenate volume (*see* **Note 2**).

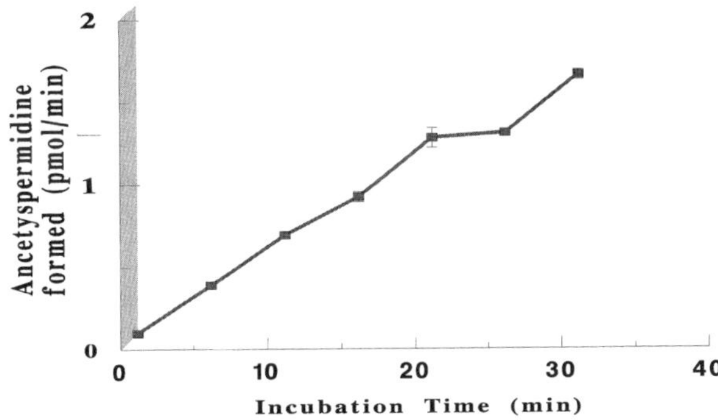

Fig. 4. Nonenzymatic acetylation of spermidine.

Table 1
Gradient Parameters for the HPLC Separation of the Polyamines

Time (min)	% Buffer A	% Buffer B
0	100	0
30	40	60
40	0	100
45	0	100
55	100	0

Note: A control tube that contains identical substrate and cofactor mixtures, but that lacks the enzyme extract, must be included in the assay procedure to enable all SAT activities to be corrected for nonenzymatic acetylation. Typical blank values were 194 ± 66 dpm. Nonenzymatic acetylation does increase with time **(Fig. 4)**.

3.3. Separation of the Individual Acetylspermidine Products

The separation and identification of the products of the acetyltransferase assay employed the use of HPLC with radiomatic detection. This method is a modification of the HPLC method developed by Seiler and Knodgen ***(21)***, as described by Wallace et al. ***(3)***. Briefly, the polyamine separation is achieved by ion-pair reverse-phase chromatography employing gradient elution (shown in **Table 1**) of acid extracts at a flow rate of 1.5 mL/min. The column eluent containing the separated polyamines then undergoes postcolumn derivatization on mixing with 3-phalaldehyde-2-mercaptoethanol reagent in borate buffer flowing at 1.5 mL/min at a ratio of 1:1. The derivatizing reagent attaches a fluorescent moiety to the primary amino group on the polyamine molecule, resulting in a fluorescent product, thus enabling the sensitive detection and

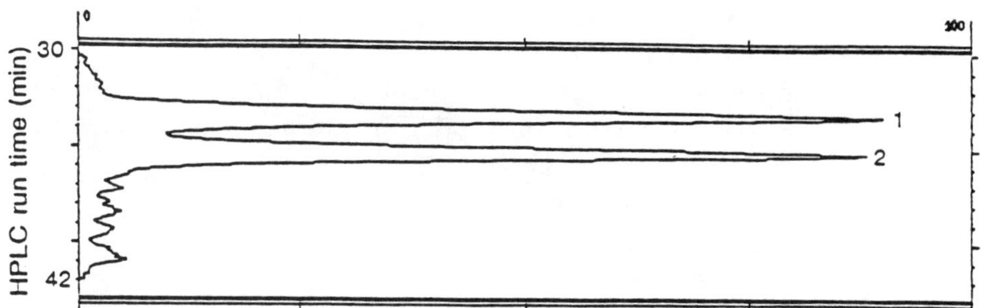

Fig. 5. Identification of *N*-acetylspermidine derivatives by HPLC. Peaks are identified as: 1: N^1-acetylspermidine; 2: N^8-acetylspermidine. The remainder of the sample was pooled and 100 µL was injected onto the HPLC as described previously *(3)*. Individual peaks were identified using external standard quantification with authentic N^1- and N^8-acetylspermidine.

quantification of the separated polyamines by fluorescence spectrometry. In addition, the 3 mL/min eluting from the fluorimeter outlet is mixed continuously with Pico-fluor 40 scintillation fluid (7 mL/min) as they pass through the radiomatic detector flow cell. The peaks on the radiomatic detector trace from the Flo-1β detector **(Fig. 5)** were easily identified as N^1 and N^8-acetylspermidine according to their retention times relative to those obtained from the fluorescence trace (*see* **Fig. 5**).

1. Pool the remaining 60 µL of the reaction products from each triplicate assay, dilute (1:1) with 0.2*M* $HClO_4$, and centrifuge at 12,000g_{av} for 2 min.
2. Inject 100 µL of this preparation onto the HPLC gradient system described previously.
3. The proportion of each isomer allows the estimation of the N^1- and N^8-SAT activity from the total SAT activity and is expressed as pmol of product formed/min/mg of protein.

4. Notes

1. In preparing the assay: To cut down on pipeting small volumes it may be more convenient to make up an assay mix. Calculate the number of assays to be done, add 2, and prepare a mix containing enough of the buffer and the substrate (spermidine or spermine) for the calculated number of assays. Then pipet 20 µL of the mix into each tube. For example: if you have 20 assays to carry out prepare enough assay mix for 22 by multiplying the amount of buffer required per assay by 22:

$$10\ \mu L \times 22 = 220\ \mu L \quad (1)$$

Then do the same calculation for the substrate.

2. To improve reproducibility in the assay ensure there is sufficient protein present. For example, if you are exposing cells to toxic agents that will kill the cells, you will need to increase the number of plates or increase the amount of tissue used in the preparation of the cytosolic extract.

References

1. Matsui, I. and Pegg, A. E. (1980) Increase in acetylation of spermidine in rat liver extracts brought about by treatment with carbon tetrachloride. *Biochem. Biophys. Res. Commun.* **92,** 1009–1015.
2. Casero, R. A. and Pegg, A. E. (1993) Spermidine/spermine N^1-acetyltransferase—the turning point in polyamine metabolism. *FASEB J.* **7,** 653–661.
3. Wallace, H. M., Nuttall, M. E., and Robinson, F. C. (1988) Acetylation of spermidine and methylglyoxal bis(guanylhydrazone) in baby hamster kidney cells (BHK-21/C13). *Biochem. J.* **253,** 223–227.
4. Della-Ragione, F. and Pegg, A. E. (1983) Studies on the specificity and kinetics of rat liver spermidine/spermine N^1-acetyltransferase. *Biochem. J.* **213,** 701–706.
5. Seiler, N. (1987) Functions of polyamine acetylation. *Can. J. Physiol. Pharmacol.* **65,** 2024–2035.
6. Wallace, H. M. and Keir, H. M. (1981) Uptake and excretion of polyamines from baby hamster kidney cells (BHK-21/C13): the effect of serum on confluent cell cultures. *Biochim. Biophys. Acta* **676,** 25–30.
7. Wallace, H. M. (1987) Polyamine catabolism in mammalian cells: excretion and acetylation. *Med. Sci. Res.* **15,** 1437–1440.
8. Pegg, A. E., Pakala, R., and Bergeron, R. J. (1990) Induction of spermidine/spermine N^1-acetyltransferase activity in Chinese hamster ovary cells by N^1, N^{12} bis(ethyl)norspermidine and related compounds. *Biochem. J.* **267,** 331–338.
9. Melvin, M. A. L. and Keir, H. M. (1978) Polyamine metabolism in BHK-21/C13 cells. Loss of spermidine from cells following transfer to serum-depleted medium. *Exper. Cell Res.* **111,** 231–236.
10. Mackarel, A. J. and Wallace H. M. (1992) Effect of 5-fluorouracil on polyamine excretion in human colonic cancer cells. *Biochem. Soc. Trans.* **21,** 50S.
11. Brunton, V. G., Grant, M. H., and Wallace, H. M. (1990) Spermine toxicity and glutathione depletion in BHK-21/C13 cells. *Biochem. Pharmacol.* **40,** 1893–1900.
12. Abdel-Monen, M. M., Ohno, K., Fortuny, I. E., and Theologides, A. (1975) Acetylspermidines in human urine. *Lancet* **ii,** 1210.
13. Loser, C., Folsch, U. R., Paprontу, C., and Creutzfeld, W. (1990) Polyamines in colorectal cancer. Evaluation of polyamine concentrations in colon tissue, serum and urine of 50 patients with colorectal cancer. *Cancer* **65,** 958–966.
14. Russell, D. H., Levy, C. C., Schimpff, S. C., and Hawk, I. A. (1971) Urinary polyamines in cancer patients. *Cancer Res.* **31,** 1555–1558.
15. Lipton, A., Sheehan, L. M., and Kessler, G. F. (1975) Urinary polyamine levels in human cancer. *Cancer* **35,** 464–468.
16. Seiler, N., Bolkenius, F. N., and Knodgen, B. (1985) The influence of catabolic reactions on polyamine excretion. *Biochem. J.* **225,** 219–226.

17. Poulin, R., Wechter, R. S., and Pegg, A. E. (1991) An early enlargement of the putrescine pool is required for growth of L1210 mouse leukaemia cells under hypoosmotic stress. *J. Biol. Chem.* **266,** 6142–6151.
18. Poulin, R., Coward, J. K., Lakanen, L. R., and Pegg, A. E. (1993) Enhancement of the spermidine uptake system and lethal effects of spermidine overaccumulation in ornithine decarboxylase overproducing L1210 cells under hypoosmotic stress. *J. Biol. Chem.* **268,** 4690–4698.
19. Libby, P. R. (1978) Calf liver nuclear N-acetyltransferases. Purification and properties of two enzymes with both spermidine acetyltransferase and histone acetyltransferase activities. *J. Biol. Chem.* **253,** 233–237.
20. Lowry, O. H., Rosebrough, N. J., Farr, A. L., and Randall, R. J. (1951) Protein measurement with the Folin phenol reagent. *J. Biol. Chem.* **193,** 265–275.
21. Seiler, N. and Knodgen, B. (1980) High performance liquid chromatographic procedure for the simultaneous determination of the natural polyamines and their monoacetyl derivatives. *J. Chromatog.* **221,** 227–235.

7

Assay of Spermidine N^8-Acetyltransferase

Jim Blankenship

1. Introduction

Spermidine undergoes three different metabolic reactions in mammalian tissues. It can be converted to spermine by the addition of an aminopropyl group, or it can be acetylated on either of the terminal nitrogens using acetyl-coenzyme A as the acetyl donor *(1)*. Acetylation on the terminal nitrogen adjacent to the 3-carbon chain is termed N^1-acetylation and is catalyzed by a cytoplasmic enzyme, spermidine/spermine N^1-acetyltransferase *(2)*. This N^1-acetylspermidine is converted by polyamine oxidase to putrescine, the normal precursor for spermidine. Thus, N^1-acetylation serves as the initial step in the conversion of spermidine to putrescine.

Acetylation on the terminal nitrogen adjacent to the 4-carbon chain produces N^8-acetylspermidine and is catalyzed by spermidine N^8-acetyltransferase *(3,4)*. This is a nuclear enzyme associated with chromatin with characteristics similar to enzymes involved in histone acetylation, but apparently under different regulation *(5)*. N^8-acetylspermidine is not a substrate for polyamine oxidase and thus is not converted to putrescine, as is N^1-acetylspermidine *(1)*. The enzyme apparently responsible for metabolism of N^8-acetylspermidine is N^8-acetylspermidine deacetylase, which converts this acetylated compound back to spermidine (*6,7*; *see* Chapter 8).

Because of the similarities of substrates (acetylCoA and spermidine) and products, assays of spermidine N^8-acetyltransferase require that the enzyme source used in the assay does not contain spermidine/spermine N^1-acetyltransferase in order to avoid measuring both of these activities in the same assay. Since spermidine N^8-acetyltransferase is a nuclear enzyme and spermidine/spermine N^1-acetyltransferase is contained in the cytoplasm, the simplest method for separating these enzymes and obtaining spermidine N^8-acetyl-

From: *Methods in Molecular Biology, Vol. 79: Polyamine Protocols*
Edited by: D. Morgan Humana Press Inc., Totowa, NJ

transferase in a form suitable for this assay is to isolate nuclei and prepare a chromatin or nuclear extract *(8)*. For our assays we use a simple one-step purification of nuclei *(3)*. These isolated nuclei can be resuspended and used directly in the assay, or for a more active preparation, nuclei can be disrupted and a simple chromatin preparation obtained as the enzyme source for the assay.

2. Materials

2.1. Materials for Isolation of Nuclei

1. Animals: adult rats, 150–250 g.
2. 0.25*M* Sucrose–TKM buffer: 0.25*M* sucrose, 0.05*M* Tris-HCl, pH 7.5, 0.025*M* KCl, 0.005*M* $MgCl_2$ (store at 0–4°C).
3. 30 mL glass/Teflon pestle (Potter/Elvehjem, 0.004–0.006 in. clearance) tissue grinder and an appropriate motor for power homogenization (Wheaton #358049, Fisher Scientific, Pittsburgh, PA).
4. Cheesecloth.
5. 2.1*M* Sucrose-TKM buffer. 2.1*M* sucrose, 0.05*M* Tris-HCl, pH 7.5, 0.025*M* KCl, 0.005*M* $MgCl_2$ (store at 0–4°C)
6. SW 25.2 rotor (and appropriate tubes) for Beckman ultracentrifuge or any swinging bucket rotor and centrifuge capable of achieving 100,000*g* for 1 h.

2.2. Materials for Preparation of Chromatin

1. 30 mL glass/Teflon pestle (Potter/Elvehjem, 0.004–0.006 in. clearance) tissue grinder (Wheaton #358049, Fisher Scientific).
2. 0.08*M* NaCl-0.02*M* EDTA, pH 6.3.
3. Centrifuge capable of 10,000*g* for 10 min, 4–5 times.
4. 0.05*M* Tris-HCl, pH 7.8.

2.3. Materials for Assay Procedure

1. 1.5 mL Eppendorf tubes.
2. Freshly prepared chromatin (100 μg protein/50 μL).
3. 0.05*M* Tris-HCl, pH 7.8.
4. 5.0 m*M* spermidine made up in 0.05*M* Tris-HCl, pH 7.8.
5. 20 μ*M* [1-^{14}C]acetyl-CoA (40–60 mCi/mmol) made up in 0.05*M* Tris-HCl, pH 6.0 (*see* **Note 3**). This will provide approx 100,000 dpm/5 μL aliquot added to each assay tube. It is important that acetylCoA be stored in solution at an acidic pH to prevent degradation. It can be stored at –20°C in aliquots for several months.
6. 12*M* hydrochloric acid (concentrated HCl).
7. Shaking water bath for incubating assay tubes at 37°C.
8. Microcentrifuge.
9. 12 × 75-mm disposable glass tubes.
10. Dowex 50W X8-100 ion exchange resin washed according to the following procedure. To 100 g of resin, add 500 mL of 1.0*M* HCl and stir for 5 min. Allow the resin to settle for at least 30 min and remove (decant) and discard the supernatant.

Repeat these wash steps with 1.0*M* HCl three times and during the last wash transfer the stirred resin mixture to a liter-graduated cylinder to measure the settled volume. Allow the resin to settle again for at least 30 min and remove (decant) and discard the supernatant. Add a volume of water equal to the settled volume of the resin and store this washed resin at 0–4°C for later use. This resin can be stored for up to 6 mo.

11. Vortex mixer.
12. General purpose (3000 rpm) centrifuge with rotor for 12 × 75 mm disposable glass tubes (example-IEC, model K, Fisher Scientific).
13. 1.0*M* HCl.
14. 1.5*M* NaOH.
15. Scintillation vials, scintillation cocktail, and liquid scintillation spectrometer.

3. Methods

3.1. Isolation of Nuclei (see Note 1)

1. Sacrifice rat by cervical dislocation.
2. Immediately remove liver and place in ice-cold 0.25*M* sucrose-TKM buffer. All the following steps of enzyme preparation should be done at 0–4°C.
3. Mince 5.0 g liver and homogenize in 10 mL 0.25*M* sucrose-TKM buffer using 12 complete strokes of a 30 mL or larger glass-Teflon tissue grinder.
4. Filter the homogenate through four layers of cheesecloth.
5. Mix the filtered homogenate with 2 vol (approx 30 mL) of 2.1*M* sucrose-TKM buffer.
6. Carefully layer this mixture over a 6 mL voluume (cushion) of 2.1*M* sucrose-TKM buffer in each of three centrifuge tubes for SW 25.2 rotor. The volume of the mixture should be distributed equally among the three tubes. If the tubes are not filled within 5 mm of the top, carefully layer sufficient 0.25*M* sucrose-TKM buffer to fill the tubes and thus prevent collapse of the tubes during centrifugation.
7. Centrifuge the tubes at 25,000 rpm (107,000*g*) for 1 h at 4°C.
8. Remove the supernatant (including cushion) and discard. Wipe the walls of the tubes with tissues.
9. The small, translucent pellets contain purified nuclei. For highest activity in the spermidine N^8-acetyltransferase assay, nuclear pellets are taken through the chromatin preparation procedure.

3.2. Preparation of Chromatin (see Note 2)

1. Suspend the three nuclear pellets from above in 15 mL of 0.08*M* NaCl-0.02*M* EDTA, pH 6.3, with gentle homogenization by hand in a 30-mL glass-Teflon tissue grinder.
2. Centrifuge at 10,000*g* for 10 min and discard supernatant.
3. Using the pellet from step 2, repeat above steps (1 and 2) twice with 0.08*M* NaCl-0.02*M* EDTA, pH 6.3.
4. Using the pellet from step 3, repeat steps 1 and 2 with resuspension in 0.05*M* Tris-HCl, pH 7.8, followed by centrifugation.

5. The final chromatin preparation is resuspended in a small volume (5 mL or less) of 0.05*M* Tris-HCl, pH 7.8. The volume chosen for this final resuspension should yield approx 100 µg of protein/50 µL of solution. This enzyme preparation retains activity for up to 2 wk on storage at –70°C.

3.3. Assay Procedure

1. Assays are run in 1.5 mL Eppendorf tubes at a final concentration of 1.0 m*M* spermidine, 1 µ*M* [1-^{14}C]acetylCoA, and chromatin containing about 100 µg of protein in 0.05*M* Tris-HCl, pH 7.8, with a total volume of 100 µL.
2. To each vial, add 20 µL of 5 m*M* spermidine in 0.05*M* Tris-HCl, pH 7.8.
3. To each vial, add 50 µL of freshly prepared chromatin (about 100 µg of protein) in 0.05*M* Tris-HCl, pH 7.8.
4. To each vial, add 25 µL 0.05*M* Tris-HCl, pH 7.8.
5. Following preincubation for 3 min at 37°C, the reactions are initiated by adding 5 µL of 20 µ*M* [1-^{14}C]acetylCoA in 0.05*M* Tris-HCl, pH 6.0 (*see* **Note 3**).
6. Incubations are carried out with shaking for 10 min at 37°C.
7. Reactions are terminated by the addition of 20 µL of 12*M* HCl and vortexing.
8. After sitting on ice for 5 min, vials are again vortexed and centrifuged for 5 min at maximum speed in a microcentrifuge.
9. Transfer 90 µL of the supernatant to 12 × 75-mm disposable glass tubes containing 0.5 mL of the washed Dowex 50W ion exchange resin (1.25 mEq/0.5 mL). These tubes were prepared prior to the assay by pipeting 0.5 mL vol from a resin suspension being stirred continuously. This procedure provided 0.25 mL of packed resin to each tube (*see* **Note 4**).
10. Add 1 mL of 1*M* HCl to each tube and vortex twice over a 5 min period. Centrifuge for 3 min at low speed (3000 rpm) in a general purpose centrifuge.
11. Remove (by aspiration) and discard the supernatant.
12. Wash the resin three times with 2 mL of water, each wash followed by vortexing and centrifugation as previously described (**steps 10** and **11**).
13. At the end of the water washes, the reaction product is extracted from the resin by adding 0.75 mL of 1.5*M* NaOH to each tube.
14. Vortex each tube for 30 s twice over a 5 min period and centrifuge as above (**step 10**).
15. Transfer 0.5 mL of the supernatant to scintillation vials, add scintillation cocktail, and measure radioactivity by liquid scintillation spectrometry.
16. Enzyme activity is usually reported as millimoles of product (N^8-acetylspermidine) formed per milligram of protein per 10 min. The millimoles of product formed in 10 min can be calculated from

$$[(\text{dpm of radioactivity measured in the assay}) / (2.22 \times 10^6 \text{ dpm}/\mu\text{Ci})] / [\text{specific activity in }\mu\text{Ci/mmol of }[1\text{-}^{14}\text{C}]\text{acetylCoA}] \quad (1)$$

The final result is calculated as (mmol of product formed in 10 min)/(mg of protein in the assay).

4. Notes

1. For meaningful assays of spermidine N^8-acetyltransferase, it is essential that the enzyme be separated from spermidine/spermine N^1-acetyltransferase or else a mixture of N^1- and N^8-acetylated spermidine will be produced. Both of these compounds will be detected by the procedures described in **Subheading 3.1.** The isolation of nuclei, chromatin, or a purified enzyme are the only solutions presently available for this problem because there are no selective inhibitors available for either enzyme.
2. During the chromatin preparation nuclei are ruptured and with each wash step the chromatin will swell and become more soluble. In some chromatin preparations, the swelling/solubilization proceeds more rapidly and it does not produce an intact pellet with centrifugation. Under this circumstance, chromatin will be lost if the procedure is continued unchanged. Best yields will be obtained if some of the intermediate wash steps are eliminated.
3. It is important to avoid diluting and/or storing [1-^{14}C]acetylCoA in solutions with a pH above 7.0 to avoid breakdown and loss of acetylating capacity.
4. An effective alternative to the use of Dowex resin for recovery of the radiolabeled product involves the use of cellulose phosphate paper disks *(2)*. This method is commonly used in assays of spermidine/spermine N^1-acetyltransferase and can readily be adapted to spermidine N^8-acetyltransferase assays without significant changes.

References

1. Seiler, N. (1987) Functions of polyamine acetylation. *Can. J. Physiol. Pharmacol.* **65,** 2024–2035.
2. Della Ragione, F. and Pegg, A. E. (1982) Purification and characterization of spermidine/spermine N^1-acetyltransferase from rat liver. *Biochemistry* **21,** 6152–6158.
3. Blankenship, J. and Walle, T. (1977) Acetylation of spermidine and spermine by rat liver and kidney chromatin. *Arch. Biochem. Biophys.* **179,** 235–242.
4. Libby, P. R. (1978) Calf liver nuclear N-acetyltransferases: purification and properties of two enzymes with both spermidine acetyltransferase and histone acetyltransferase activities. *J. Biol. Chem.* **253,** 233–237.
5. Desiderio, M. A., Mattel, S., Biondi, G., and Colombo, M. P. (1993) Cytosolic and nuclear spermidine acetyltransferases in growing NIH 3T3 fibroblasts stimulated with serum or polyamines: relationship to polyamine-biosynthetic decarboxylases and histone acetyltransferase. *Biochem. J.* **293,** 475–479.
6. Blankenship, J. (1978) Deacetylation of N^8-acetylspermidine by subcellular fractions of rat tissue. *Arch. Biochem. Biophys.* **189,** 20–27.
7. Blankenship, J. and Marchant, P. (1984) Metabolism of N^1-acetylspermidine and N^8-acetylspermidine in rats. *Proc. Soc. Exper. Biol. Med.* **177,** 180–187.
8. Blankenship, J. and Walle, T. (1978) In vitro studies of enzymatic synthesis and metabolism of N-acetylated polyamines, in *Advances in Polyamine Research*, vol. 2 (Campbell, R. A., et al., eds.), Raven, New York, pp. 97–110.

III

Assay Methods for Enzymes of Polyamine Catabolism

8

Assay of N^8-Acetylspermidine Deacetylase

Jim Blankenship

1. Introduction

N^8-Acetylspermidine deacetylase is a cytoplasmic enzyme found in a wide variety of tissues in higher organisms *(1)*. This enzyme catalyzes the deacetylation of polyamines *N*-acetylated on the terminal amino group on the 4-carbon end of the molecule. Thus, N^8-acetylspermidine and *N*-acetylputrescine are substrates for this enzyme, whereas polyamines acetylated on aminopropyl groups, such as N^1-acetylspermidine and *N*-acetylspermine, are not substrates *(2)*. The metabolism of these latter molecules, acetylated on aminopropyl groups, is catalyzed by polyamine oxidase and does not involve a simple deacetylation, but instead there is an oxidative cleavage of the entire acetylated aminopropyl group to yield putrescine or spermidine *(3,4)*.

N^8-Acetylspermidine deacetylase is assayed by a fairly simple procedure, but does require the chemical synthesis of the radiolabeled substrate, N^8-[acetyl-^{3}H]acetylspermidine *(1)*. The assay involves cleavage of the radiolabeled acetyl group by the enzyme from rat liver, extraction of the free ^{3}H-acetic acid with ethyl acetate, and measurement of radioactivity by liquid scintillation spectrometry.

2. Materials

2.1. Materials for Synthesis and Purification of Substrate

1. Spermidine (free base).
2. 5 mCi total, acetic anhydride, [^{3}H]-($CH_3CO)_2O$, 50 mCi/mmol.
3. Amberlite cation exchange resin CG-50 (100–200 mesh) prepared by washing alternatively with 20 vol 3*M* HCl and 3*M* NH_4OH. This washing sequence should be repeated three times followed by a series of washings with deionized water until the pH reaches 9.5.

From: *Methods in Molecular Biology, Vol. 79: Polyamine Protocols*
Edited by: D. Morgan Humana Press Inc., Totowa, NJ

4. Chromatography column, 25 × 1.5 cm, filled with the aforementioned prepared Amberlite CG-50 to a level of 20 cm.
5. Fraction collector.
6. Lyophylizer.
7. Liquid scintillation spectrometer.

2.2. Materials for Assay

1. Animals: adult rats 150–250 g.
2. SPM buffer: 0.25*M* sucrose, 0.05*M* NaH_2PO_4 (pH 7.4), 0.005*M* $MgCl_2$.
3. Polytron homogenizer, model PCU-2 (Brinkmann Instruments, Westbury, NY).
4. Ultracentrifuge and rotor capable of 105,000*g* for 1 h.
5. Assay tubes: heavy-duty glass plain conical centrifuge tubes (13 mL capacity) with flat head glass stoppers (Kimble # 4517413, Fisher Scientific, Pittsburgh, PA).
6. Shaking water bath for incubating assay tubes at 37°C.
7. Eberbach 6000 variable speed shaker (Fisher Scientific, Pittsburgh, PA).
8. Scintillation vials, scintillation cocktail, and liquid scintillation spectrometer.

3. Methods

3.1. Synthesis and Purification of Substrate (see Note 1)

1. To 1 mL of acetonitrile containing 1 μmol of spermidine (free base), add 0.1 μmol of ^{3}H-labeled acetic anhydride. For this reaction, maintain the mixture in an acetone dry-ice bath for 1 h and allow to stand at room temperature overnight. Remove the acetonitrile by evaporation under a stream of nitrogen and dissolve the residue in 10 mL water.
2. Load the 10 mL of water containing the reaction products on the Amberlite-filled column and wash the column with 50 mL of deionized water.
3. Wash the column with 100 mL of 0.5*M* NH_4OH to elute any diacetylspermidine.
4. Begin elution of monoacetylated spermidines with addition of 150 mL of 2.0*M* NH_4OH at a rate of 2.0 mL/min and collection of 2.0 mL fractions. The order of elution is N^8-acetylspermidine (approx fractions 16–25) first followed by N^1-acetylspermidine (approx fractions 39–49).
5. Analyze each fraction for content of radioactivity and confirm identity of acetylspermidine derivatives by HPLC or TLC. A simple thin-layer chromatographic method involves separation on silica gel plates (type 60F-254, 0.25 mm thickness, Brinkmann) with a solvent system of chloroform:methanol:NH_4OH (2:2:1). The polyamines are detected by measuring radioactivity and by spraying with ninhydrin (20 mg/mL in butanol) and developing color at 110°C.
6. Fractions containing only N^8-acetylspermidine are pooled and the water and NH_4OH are removed by lyophilization.
7. The white powdery residue is dissolved in deionized water and the pH was adjusted to 2.0 with 2*M* HCl to yield the hydrochloride salt of N^8-[acetyl-^{3}H]acetylspermidine.

8. Adjust the concentration of N^8-[acetyl-^{3}H]acetylspermidine by dilution or addition of unlabeled compound to yield a concentration of approx 2.5 nmol (25,000 dpm)/ 10 μL of solution.

3.2. Source of Enzyme

1. Sacrifice rat by cervical dislocation.
2. Immediately remove liver and place in ice-cold SPM buffer. All the following steps of enzyme preparation should be done at 0–4°C.
3. Mince 3.0 g liver and homogenize in 12 mL SPM buffer for 15 s using a Brinkmann Polytron homogenizer at the maximum setting.
4. Centrifuge the homogenate 105,000*g* for 60 min.
5. Remove the supernatant (cytosol) fraction and store at 0–4°C for use as the enzyme source for the N^8-acetylspermidine deacetylase.

3.3. Assay Procedure

1. Each assay mixture should consist of 0.125 mmol sucrose, 2.5 μmol $MgCl_2$ and 25 μmol NaH_2PO_4, pH 7.4, 2.5–12.5 nmol N^8-[acetyl-^{3}H]acetylspermidine, and enzyme (supernatant fraction) in a total volume of 0.5 mL.
2. Each assay is run in a conical centrifuge tube that can be sealed with glass stoppers, but the tubes are left uncovered until the extraction step. Assays mixtures are prepared and maintained at 0–4°C prior to incubation.
3. To each assay tube, add 10–50 μL of N^8-[acetyl-^{3}H]acetylspermidine (2.5–12.5 nmol and 25,000–125,000 dpm).
4. To each assay tube, add 390 (or less) μL of SPM buffer. The volume of SPM buffer is adjusted to provide a final volume of 500 μL based on the amounts of the other two components. This solution may contain potential inhibitors or other compounds (*see* **Note 2**).
6. To each assay tube, add 100–250 μL of enzyme preparation (supernatant fraction). This is added last to start the reaction, and the assay tubes are immediately transferred to the water bath.
7. Incubate assays by shaking tubes for 10 min at 37°C.
8. Stop assays by transferring tubes to an ice bath and immediately adding 500 μL of 1.0*M* HCl–0.005*M* acetic acid.
9. Add 3 mL of ethyl acetate to each tube and seal with glass stoppers.
10. Shake the tubes vigorously for 1 h.
11. Centrifuge the tubes for 5 min at 1500*g* in a swinging bucket rotor.
12. Transfer (by pipet) 2 mL of the upper (ethyl acetate) layer to scintillation vials, add scintillation cocktail, and measure radioactivity by liquid scintillation spectrometry.
13. Results of assays are usually reported as dpm of radioactivity extracted into ethyl acetate/mg of protein/10 min. This can be converted to millimoles of ^{14}C-acetate released if the specific radioactivity of the substrate is known. The dpm of radioactivity released in 10 min can be calculated from

$$\text{(dpm of radioactivity extracted in the assay)} \times 1.5 \qquad (2)$$

to compensate for loss during extraction and transfer. The final result is calculated from (dpm released in 10 min)/(mg of protein in the assay).

4. Notes

1. In assaying for N^8-acetylspermidine deacetylase activity, it is essential that the N^8-acetylspermidine used as substrate not be contaminated with N^1-acetylspermidine ***(5)***. Such contamination raises the possibility that polyamine oxidase activity may be mistakenly measured as part of the deacetylase reaction. In fact in this situation, polyamine oxidase does not deacetylate either N^8-acetylspermidine or N^1-acetylspermidine, but rather by oxidative deamidation releases *N*-acetylpropionaldehyde from only N^1-acetylspermidine. This reaction can lead to confusion with the deacetylation reaction because the radiolabeled *N*-acetylpropionaldehyde is extracted by ethyl acetate, just as is the acetic acid. Thus, it is necessary to confirm that the substrate, whether synthesized in the laboratory or obtained commercially, is pure N^8-acetylspermidine without N^1-acetylspermidine contamination. This confirmation can follow the chromatographic purification methods described in **Subheading 3.1.** or HPLC.
2. Inhibition studies are conducted by dissolving potential inhibitors in SPM buffer and adding as a fraction of the SPM buffer component in the assay mixture. Two compounds, 7-[*N*-(3-aminopropyl)amino]heptan-2-one ***(6)*** and 7-amino-2-heptanone ***(7,8)***, have been reported to competitively inhibit N^8-acetylspermidine deacetylase and are useful for inhibition studies in this assay system.

References

1. Blankenship, J. (1978) Deacetylation of N^8-acetylspermidine by subcellular fractions of rat tissue. *Arch. Biochem. Biophys.* **189,** 20–27.
2. Seiler, N. (1987) Functions of polyamine acetylation. *Can. J. Physiol. Pharmacol.* **65,** 2024–2035.
3. Holtta, E. (1977) Oxidation of spermidine and spermine in rat liver: purification and properties of polyamine oxidase. *Biochemistry* **16,** 91–100.
4. Blankenship, J. and Marchant, P. (1984) Metabolism of N^1-acetylspermidine and N^8-acetylspermidine in rats. *Proc. Soc. Exper. Biol. Med.* **177,** 180–187.
5. Marchant, P., Manneh, V., and Blankenship, J. (1986) N^1-acetylspermidine is not a substrate for N-acetylspermidine deacetylase. *Bioch. Biophys. Acta* **881,** 297–299.
6. Marchant, P., Dredar, S., Manneh, V., Al Shabanah, O., Matthews, H., Fries, D., and Blankenship, J. (1989) A selective inhibitor of N^8-acetylspermidine deacetylation in mice and HeLa cells without effects on histone deacetylation. *Arch. Biochem. Biophys.* **273,** 128–136.
7. Desiderio, M. A., Bernhardt, A., and Mamont, P. S. (1991) Heterogeneity of rat liver nuclear spermidine N^8- and histone acetyltransferases. *Life Chem. Rep.* **9,** 57–63.
8. Desiderio, M. A., Weibel, M., and Mamont, P. S. (1992) Spermidine nuclear acetylation in rat hepatocytes and in logarithmically growing rat hepatoma cells: comparison with histone acetylation. *Exper. Cell Res.* **202,** 501–506.

9

Hydrogen Peroxide Assay for Amine Oxidase Activity

R. James Storer and Antonio Ferrante

1. Introduction

Amine oxidase activity can be assessed by measuring the conversion of substrate to product, or by measuring the formation of concomitantly formed hydrogen peroxide. Fluorometry offers greater sensitivity and selectivity than conventional colorimetric methods for measuring hydrogen peroxide, for example those based on coupled oxidation of chromogenic compounds with peroxidase, such as *o*-dianisidine ***(1–4)*** or indigo disulfonate ***(5)***. The *o*-dianisidine method is not specific, is sensitive to superoxide radicals ***(6)***, and *o*-dianisidine may also act as an inhibitor of enzyme activity ***(7)***. The fluorometric method permits the use of a variety of substrates and continuous monitoring of the enzyme reaction, which are important for kinetic studies. Radiometric methods are limited by the available substrates and the measurement of product formation at a single time-point. Measurement of oxygen consumption, either manometrically or using a Clark electrode (polargraphic method), allow the use of multiple substrates, but are relatively insensitive and technically difficult. Coupled ammonia measurements are also relatively insensitive ***(8)*** and limited to amine oxidases acting at the primary amino groups. The method described in this chapter is comparable in sensitivity to the radiometric methods ***(9,10)*** and is relatively simple to perform.

1.1. Fluorometric Amine Oxidase Assay

The hydrogen peroxide formed in the amine oxidase reaction **(Fig. 1A)** is measured fluorometrically by coupling it to the oxidation of homovanillic acid (I) to the fluorescent biphenyl structure (HVA dimer, II; **Fig. 1B**) in the presence of horseradish peroxidase (donor: H_2O_2-oxidoreductase; HRPO; EC 1.11.1.7) ***(11)*** using a modification of the method first described by Guilbault

From: *Methods in Molecular Biology, Vol. 79: Polyamine Protocols*
Edited by: D. Morgan Humana Press Inc., Totowa, NJ

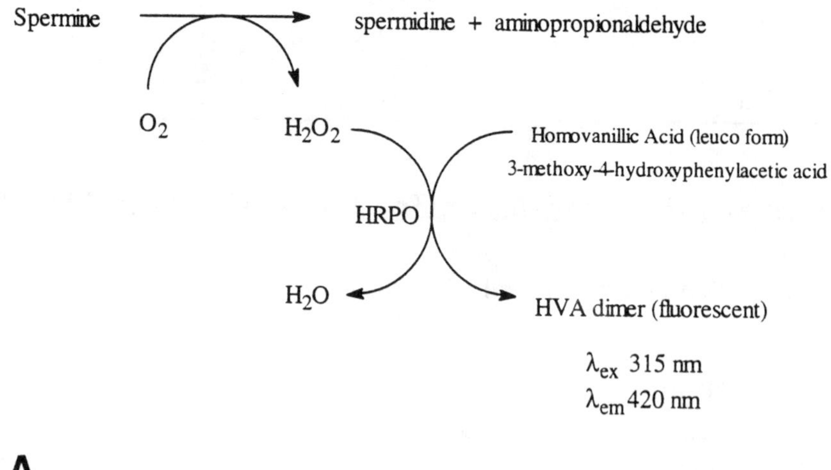

Fig. 1. **(A)** Schematic for the coupled enzyme assay of spermine oxidation by amine oxidase. **(B)** Oxidation of homovanillic acid by the peroxidase(HRPO)-H_2O_2 complex.

et al. ***(12,13)*** and Snyder and Hendley ***(14,15)***. Homovanillic acid (HVA) can act as an electron acceptor with peroxidase (hydrogen donor). HVA solutions are stable, an advantage over the diacetylfluorescine ***(11)*** or scopoletin (6-methoxy-7-hydroxy-1,2-benzopyrone, 7-hydroxy-6-methoxycoumarin) methods ***(16)***. Unlike more sensitive fluorogenic substrates, HVA does not inhibit polyamine oxidase ***(17,18)*** and the fluorescent dimer is more stable ***(19)***.

The HRPO reaction will underestimate H_2O_2 concentrations when hydrogen donors, such as serum albumin or reduced hemoglobin, or reducing agents, such as β-mercaptoethanol or DTT ***(20)***, hydroxyindole substrates of HRPO, such as serotonin (5-hydroxytryptamine) ***(21)***, or phenolic compounds, such as phenol red, which compete for HRPO-H_2O_2 (compound II), are present in the reaction mixture ***(17,21–26)***. If a lag phase is observed, it is likely that endogenous hydrogen donors are oxidized by the H_2O_2 from the amine oxidase reaction. On exhaustion of the hydrogen donors, the H_2O_2 then oxidizes the HVA, and H_2O_2 production rates can be measured more reliably ***(16)***.

Hydrogen peroxide would normally be destroyed by catalase. However, in the presence of low concentrations of H_2O_2 and a suitable acceptor, catalase will tend to act peroxidatically ***(27)***. The presence of peroxidase ensures a negligible reaction of H_2O_2 with catalase because of the large differences in affinities (K_m) of catalase and peroxidase for hydrogen peroxide ***(28)***.

The high velocity of the HRPO reaction with H_2O_2 and homovanillic acid is well established ***(12)*** and sufficient HRPO is included in the assay to accurately reflect H_2O_2 production by the polyamine oxidase system ***(17)***. On titrating homovanillic acid (100 μ*M*–10 m*M*) in the presence of 200 μ*M* H_2O_2 (i.e., below its rate-limiting value ***[29]***); we found that 1 mM HVA in the assay gave a maximal fluorescence response while maintaining a relatively low reagent blank. The assay described is linear for H_2O_2 concentration to 100 μ*M*. With HVA concentrations >1 m*M*, blanks are increased with no net increase in the fluorescent product. Although partial substrate inhibition of peroxidase has been observed when hydrogen peroxide concentration exceeds 5 m*M* ***(19,30)***, the peroxide generated by the amine oxidase reaction under the assay conditions described here is rapidly consumed by the peroxidase well before reaching inhibitory concentrations. The pH optimum of 7.4 was established using a radiochemical method ***(9)***.

A substrate concentration of 100 μ*M* is sufficient to prevent significant (>20%) substrate depletion during the course of the assay except, perhaps, with highly active samples, although substrate inhibition occurs and competing reactions occur when longer chain polyamines, such as spermine, are used. However, when spermine or spermidine is the substrate, it can be oxidized before the lower polyamines, which keeps the reaction linear. High substrate concentrations should be avoided because they can result in substrate inhibition or the presence of excessive concentrations of contaminant that may inhibit the enzyme being measured. Also, where substrate consumption is low the (inhibitory) effects of product accumulation are diminished ***(31)***.

When assaying amniotic fluid, Neufeld and Chayen ***(19)*** found that the activity of the sample was not proportional to sample volume; using the method of Guilbault et al. ***(12)*** they found lag times of 5–20 min and of 10–60 min using the method of Paolucci et al. ***(32)*** as a result of the presence of peroxidase inhibitors in the samples. Although with unpurified enzyme the assay results must be treated with caution (because of lag times), we have found the assay to be quite satisfactory for examining chromatography eluates where large numbers of samples must be examined in a short time in a rapid and convenient way.

This discontinuous method provides a single point determination that is suitable to screen large numbers of samples. The rate should be linear over the time of measurement and show a proportional response with increasing enzyme concentration.

1.1.2. Microplate Format

The assay can be adapted to a 96-well microtiter plate format dramatically increasing throughput.

1.1.3. Continuous Coupled Enzyme Assay

Where a more accurate determination of enzyme activity is required, such as in the determination of enzyme kinetics, the method can be adapted to continuously follow hydrogen peroxide production.

2. Materials

1. Putrescine dihydrochloride (1,4-diaminobutane), spermidine trihydrochloride (*N*-[3-aminopropyl]-1,4-butanediamine), and spermine (*N,N'*-bis[3-aminopropyl]-1,4-butanediamine) are obtained from Sigma (St. Louis, MO) or Calbiochem-Novabiochem (San Diego, CA) N^1-acetylspermine.3HCl, N^1-acetylspermidine.2HCl, histamine.2HCl and putrescine.2HCl can be obtained from Sigma.
2. High purity water (MilliQ; Millipore Corporation, Bedford, MA) is necessary for successful fluorometric assays.
3. Hydrogen peroxide solution 30% (w/v) ("100 volume" analytical reagent grade; *see* **Note 1**).
4. White pigmented 96-well flatbottomed microtiter plates (MicroFLUOR "W;" Dynatech, Chantilly, VA; *see* **Note 2**).
5. LS 50B Luminescence spectrophotometer or similar (Perkin-Elmer, Foster City, CA).
6. Spectrophotometer, UV/Visible range.
7. Water bath fitted with a shaking platform and a water recirculating immersion thermoregulator set at 37.0 ± 0.2°C.
8. Incubation buffer (10X): Dissolve Tris-HCl (28.60 g) and Tris-base (8.30 g) in MilliQ water, make up to 500 mL, filter to 0.45 μm, and store at room temperature (1X buffer = 0.05*M* Tris-HCl, pH 7.4 at 37°C; *see* **Note 3**).
9. Horseradish peroxidase (HRPO) 4000 U/mL; Type II, Sigma (*see* **Note 4**). Dissolve at 1.06 mg/mL in 1X incubation buffer.
10. Homovanillic acid (HVA) 100 m*M*: Dissolve HVA (Sigma) at 18.22 mg/mL in MilliQ water. Solution can be facilitated by gentle warming and stirring on a hot plate. Filter through a 0.45 μm membrane and store as 2 mL aliquots at –15°C (*see* **Note 5**).
11. Substrates, 10 m*M* stock solutions: Dissolve the following quantities of polyamine in MilliQ water and store at 4°C.

Substrate	Mol wt	Per 10.0 mL
Putrescine · 2HCl	161.1	1.61 mg
Spermidine · 3HCl	254.6	2.55 mg
Spermine · 4HCl	348.2	3.48 mg

12. Prepare the following reaction mixtures:

	For one assay, µL	Final concentration	Per 100 assays, mL
Buffered HRPO solution			
0.5*M* Tris-HCl (10X stock)	100	0.05*M*	10.0
HRPO (4000 U/mL stock)	10	40 U/mL	1.0
MilliQ water	590	—	59.0
Total	700		70.0
Substrate solution			
HVA (100 m*M*; *see* **Note 4**)	10	1.0 m*M*	1.0
Substrate (10 m*M*)	10	100 µ*M*	1.0
Water	180	—	18.0
	200		20.0

13. Hydrogen peroxide standards: Prepare a fresh 10 m*M* stock on each assay day. Dilute approx 97 µL of 30% (w/v; "100 volumes") hydrogen peroxide (*see* **Note 6**) with water to 100.00 mL in a very clean "A" grade volumetric flask. Determine the actual concentration of this solution spectrophotometrically (*see* **Note 7**) and prepare standards as follows:

Standard, nmol/mL	10 m*M* stock, µL	Water, µL	H_2O_2 in assay, nmol	Total volume, µL
100	10	990	10	1000
200	20	980	20	1000
400	40	960	40	1000
600	60	940	60	1000
800	80	920	80	1000

3. Methods

3.1. Discontinuous Method

1. Add sample plus isotonic buffer or saline (if necessary) to all assay sample tubes so the final sample volume will be 100 µL. Samples are added to each tube (conveniently using a Gilson Microman M25 positive displacement pipet [Gilson Medical Electronics, Villers-le-Bel, France] equipped with 3–25 or 50–250 µL capillaries and pistons).
2. Add hydrogen peroxide standards, controls, and reagent blanks (100 µL each) to further assay tubes.
3. Add buffered HRPO solution (700 µL) to each sample, standard, control, and reagent blank using an Eppendorf 4780 Multipette multipipet (Eppendorf Gerätebau Netheler + Hinz GmbH, Hamburg, Germany).

4. Preincubate the assays for 10 min at 37°C.
5. Add substrate solution (200 µL) to each sample to initiate the reaction. Incubate the assay tubes with constant shaking at approx 1 cycle/s for 30–60 min at 37.0°C (time as appropriate for the anticipated activity of the samples).
6. Quickly add 100 µL 1.0*M* NaOH stop solution to each tube using the multipipet.
7. Allow the assay tubes to equilibrate to room temperature. Then measure their fluorescence using an LS 50 luminescence spectrophotometer set to excitation λ_{315}; emission λ_{420}, with 5.0 nm slits and fitted with a no. 96 Wratten 0.6 OD neutral density attenuating filter (Kodak) on the emission window (*see* **Notes 8** and **9**).

3.2. Microplate Format

1. For each microtiter plate prepare:
   a. Buffered HRPO solution (20 mL): 2.00 mL 10X 0.5*M* Tris-HCl stock, 0.25 mL HRPO stock (4000 U/mL), and 17.75 mL water.
   b. Substrate solution (5 mL): 250 µL HVA (100 m*M* stock), 250 µL substrate (10 m*M* stock), and 4.50 mL water.
2. To each assay well add 10 µL sample and 190 µL buffered HRPO solution.
3. One "Autozero" (reagent) sample and an "autoconc" sample (usually 10 nmol H_2O_2) can be included in the assay and/or alternatively a standard curve of 0–100 nmol H_2O_2 can be set up (depending on the software running the plate reader).
4. Add substrate solution (50 µL) to all samples to initiate the reaction.
5. Incubate the plates at 37°C for an appropriate period (30–90 min).
6. Terminate the reaction by adding 25 µL 1.0*M* NaOH "stop" solution to each well.
7. Allow the plates to equilibrate to room temperature. Then measure the fluorescence using an LS 50 Luminescence spectrophotometer fitted with a plate reader: λ_{315}; λ_{420}, slits 10 nm/15 nm, "clear" emission filter.

This method is not as sensitive or accurate as the standard fluorometric assay method but 96 samples can be read in <5 min.

3.3. Continuous Coupled Enzyme Assay

The formation of the fluorophore is measured using a Perkin-Elmer LS-50B luminescence spectrophotometer equipped with a four position, water-thermostated, stirred cuvet holder. Temperature control is by means of a waterbath fitted with a Braun Thermomix 1420 water recirculating immersion thermoregulator. Reaction and control mixtures are contained in matched standard 10-mm path length rectangular quartz fluorimeter cuvets (Starna Ltd., Essex UK). The reaction mixture should contain 1.00 m*M* HVA, HRPO (2.23 purpurogallin U, 10.6 µg/mL), substrate, and sample in a total volume of 3.0 mL.

1. Place 2.86 mL 0.05*M* Tris-HCl, pH 7.4 ± 0.5 (at 37.0°C, *see* **Note 10**), 30 µL 100 m*M* HVA, and 30 µL HRPO (4000 U/mL) in a cuvet and preincubate with stirring until thermally equilibrated at 37.0 ± 0.1°C (measured in the cuvet).

2. Add the polyamine substrate (30 μL) at a concentration appropriate for kinetic data determination and incubate for a further couple of minutes to achieve thermal equilibrium.
3. Commence recording (λ_{315}; λ_{420}) to control for blank rate (according to the Fluorescence data manager software "time drive" dialog).
4. Initiate the reaction by adding enzyme preparation (50 μL, diluted with incubation buffer if necessary).
5. Monitor fluorescence at 1.0 s intervals for at least 5 min.

Fluorometric data can be related to hydrogen peroxide production by adding 0.10 and 0.50 nmol increments of standard hydrogen peroxide solution to Tris-buffered HVA/HRPO to obtain a measure of fluorescent units per nanomole of hydrogen peroxide. Under the conditions of the assay we found the maximal velocity, V_{max}, of the auxillary (coupling) enzyme, horseradish peroxidase, was 3.55 ± 0.09 μ*M*/s and = 17.3 ± 0.4 μ*M* (1.0 m*M* HVA). Thus the lag time *t* to measure 0.98*v*, the experimental velocity of the amine oxidase, was 19 s ***(33,34)***, after which time the PAO-catalyzed reaction is rate limiting. The rates of the enzyme reactions can be obtained from analysis of the slopes of the initial linear portion of progress curves of the coupled enzyme reaction. Initial velocities can be determined from the slope of a least-squares linear regression to the linear portion of the progress curve between a lag time of $t = 20$ s and $t = 80$ s using the "Obey" software (Perkin-Elmer) of the LS 50B Luminescence spectrophotometer. Substrate depletion should be <5% during this time. For each substrate concentration, the blank rate should be subtracted from the initial rate measurements to correct for the formation of the fluorophore in the absence of the oxidase.

4. Notes

1. The H_2O_2 is kept refrigerated and free of contamination to avoid any decomposition.
2. The opaque pigmentation enhances signal output, maintains a consistently low background, and prevents signal crosstalk.
3. Tris buffer should be used ***(12)*** since phosphate complexes with polyamines, which can be the cause of apparent inhibition of amine oxidases by phosphate buffers ***(6,35)***.
4. HRPO should be stored at –15°C as a desiccated powder and equilibrated to room temperature before opening. In the final assay mixture the HRPO concentration is nominally 2.23 purpurogallin U (10.6 μg/mL, or 40 μ*M*/mL). However, under the conditions of the assay it was found that the maximal velocity, V_{max}, of the HRPO was 3.55 ± 0.09 μ*M*/s (213 mU/mL), = 17.3 ± 1.4 μ*M* (at [HVA] = 1.0 m*M*), = 1.67 m*M* (at [H_2O_2] = 200 μ*M*). There is, however, some debate about the validity of K_m measurements for a hydrogen donor.
5. Homovanillic acid in aqueous stock solution is completely stable at –15°C.

6. The concentration of 30% (w/v) hydrogen peroxide stock solution can be determined by titration with ceric (IV) sulfate solution that has been standardized with anhydrous arsenous oxide ***(36)***.
7. Extinction coefficients of hydrogen peroxide, determined by us using a PYE-Unicam SP8-100 UV/Vis spectrophotometer that had been recently calibrated to the manufacturer's specifications, were ε_{230} 62.42 M^{-1}/cm and ε_{240} 39.22 M^{-1}/cm and are consistent with recent, carefully determined data from Nelson and Kiesow ***(37)***. Actual concentrations of "10 m*M*" H_2O_2 working stock solutions (about 1/1000 dilution of 10.3*M* stock ~30% (w/v) "100 vol" H_2O_2) should be checked spectrophotometrically on each occasion of use using matched quartz cuvets that have been cleaned with a suspension of magnesium oxide in water using a cotton bud, then soaked in concentrated nitric acid overnight, rinsed thoroughly with water, and dried with a stream of filtered, high purity nitrogen ***(38)***. Unless adequate precautions are taken to prevent bubbles of oxygen forming on the walls of the cuvets, serious errors can be introduced.
8. The slight increase in fluorescence of blank solutions is probably caused by oxidant impurities in the HRPO, probably Fe(III).
9. "Sipper" and "autosampler" accessories facilitate the sample reading.
10. Because the pH of Tris has a significant temperature coefficient, it is important to adjust the pH of Tris buffers to the intended pH at the temperature at which they are to be used.

References

1. Aarsen, P. N. and Kemp, A. (1964) Rapid spectrophotometric micromethod for determination of histaminase activity. *Nature* **204,** 1195.
2. Gordon, A. R. and Peters, J. H. (1967) Plasma histaminase in various mammalian species: a rapid method of assay. *Proc. Soc. Exp. Biol. Med.* **24,** 399–404.
3. Bieganski, T., Blasinska M. J., and Kusche, J. (1977) Determination of histaminase (diamine oxidase) activity by o-dianisidine test: interference of ceruloplasmin. *Agents Actions* **7,** 85–92.
4. Houen, G. and Leonardsen, L. (1992) A specific peroxidase-coupled activity stain for diamine oxidases. *Anal. Biochem.* **204,** 296–299.
5. McEwen, C. M. (1965) Human plasma monoamine oxidase. *J. Biol. Chem.* **240,** 2003–2010.
6. Kusche, J. and Lorenz, W. (1983) Diamine oxidase, in *Methods of Enzymatic Analysis* Bergmeyer, H. U., Bergmeyer, J., and Grassl, M., eds.), Verlag-Chemie, Weinheim, pp. 237–250.
7. Stoner, P. (1985) An improved spectrophotometric assay for histamine and diamine oxidase (DAO) activity. *Agents Actions* **17,** 5–9.
8. Lorenz, W., Kusche J., Hahn H., and Werle, E. (1968) Determination of the activity of diamine oxidases, urease and histidine ammonia-lyase by enzymatic assay of ammonia. *Z. Analyt. Chem.* **243,** 259–263.
9. Morgan, D. M. L. and Illei, G. (1981) Radiochemical estimation of serum polyamine oxidase activity in human pregnancy. *Med. Lab. Sci.* **38,** 49–56.

10. Smith, C. H., Maghsoudloo, M., Himuna, K., and Murphy, G. M. (1991) Pitfalls in the determination of diamine oxidase activity. *Clin. Chim. Acta* **201,** 89–94.
11. Keston, A. S. and Brant, R. (1965) The fluorometric analysis of ultramicro quantities of hydrogen peroxide. *Anal. Biochem.* **11,** 1–5.
12. Guilbault, G. G., Brignac, Jr., P. J., and Zimmer, M. (1968) Homovanillic acid as a fluorometric substrate for oxidative enzymes. *Anal. Chem.* **40,** 190–196.
13. Guilbault, G. G., Kramer, D. N., and Hackley, E. (1967) A new substrate for the fluorometric determination of oxidative enzymes. *Anal. Chem.* **39,** 271.
14. Snyder, S. H. and Hendley, E. D. (1968) A simple and sensitive fluorescence assay for monoamine oxidase and diamine oxidase. *J. Pharmacol. Exp. Ther.* **163,** 386–392.
15. Snyder, S. H. and Hendley, E. D. (1971) Sensitive fluorometric and radiometric assays for monoamine oxidase and diamine oxidase. *Meth. Enzymol.* **17,** 741–746.
16. Boveris, A., E. Martino, and Stoppani, A. O. M. (1977) Evaluation of the Horseradish peroxidase-Scopoletin method for the measurement of hydrogen peroxide formation in biological systems. *Anal. Biochem.* **80,** 145–158.
17. Zaitsu, K. and Ohkura, Y. (1980) New fluorogenic substrates for horseradish peroxidase: rapid and sensitive assays for hydrogen peroxide and the peroxidase. *Anal. Biochem.* **109,** 109–113.
18. Matsumoto, T., Furuta, T., Nimura, Y., and Suzuki, O. (1984) 3-(*p*-hydroxyphenyl) propionic acid as a new fluorogenic reagent for amine oxidase assays. *Anal. Biochem.* **138,** 133–136.
19. Neufeld, E. and Chayen, R. (1979) An evaluation of three methods for the measurement of diamine oxidase (DAO) activity in amniotic fluid. *Anal. Biochem.* **96,** 411–418.
20. Kilburn, D. M. and Taylor, P. M. (1969) Effect of sulfhydryl reagents on glucose determination by the glucose oxidase method. *Anal. Biochem.* **27,** 555–558.
21. Nelson, D. R. and Huggins, A. K. (1974) Interference of 5-hydroxytryptamine in the assay of glucose by glucose oxidase:peroxidase:chromogen based methods. *Anal. Biochem.* **59,** 46–53.
22. Nelson, D. R. and Huggins, A. K. (1974) The interaction of 5-hydroxytryptamine and related hydroxyindoles with horseradish and mammalian peroxidase systems. *Biochem. Pharmacol.* **24,** 181–189.
23. Putter, J. and Becker, R. (1983) Peroxidases, in *Methods of Enzymatic Analysis, vol. III, Enzymes 1: Oxidoreductases, Transferases* (Bergmeyer, H. U., ed.), Verlag-Chemie, Weinheim, pp. 286–293.
24. Pick, E. and Mizel, D. (1981) Rapid microassays for the measurement of superoxide and hydrogen peroxide production by macrophages in culture using an automatic enzyme immunoassay reader. *J. Immun. Meth.* **46,** 211–226.
25. Danner, D. J., Brignac, P. J., Jr., Arceneaux, D., and Patel, V. (1973) The oxidation of phenol and its reaction product by horseradish peroxidase and hydrogen peroxide. *Arch. Biochem. Biophys.* **156,** 759–763.
26. Guilbault, G. G., Brignac, P. J., Jr., and Juneau, M. (1968) New substrates for the fluorometric determination of oxidative enzymes. *Anal. Chem.* **40,** 1256–1263.

27. Aebi, H. (1962) Mitochondrial structure as a controlling factor of monoamine-oxidase activity and the action of amino-oxidase-inhibitors. *Biochem. Pharmacol.* **9,** 135–140.
28. Tipton, K. F. (1969) A sensitive fluorometric assay for monoamine oxidase. *Anal. Biochem.* **28,** 318–325.
29. Bergmeyer, H. U. (1983) *Methods of Enzymatic Analysis, Volume 1. "Fundamentals"* Verlag-Chemie, Weinheim.
30. Childs, R. E. and Bardsley, W. G. (1975) The steady state kinetics of peroxidase with 2,2'-azino-di-(3-ethyl-benzthiazoline-6-sulphonic acid) as chromogen. *Biochem. J.* **145,** 93–103.
31. Seiler, N. and Knödgen, B. (1983) *N*-(3-aminopropyl)pyrrolidin-2-one: a physiological excretory product deriving from spermidine. *Int. J. Biochem.* **15,** 907–915.
32. Paolucci, F., Cronenberger, L., and Pacheco, H. (1971) Amélioration de la méthode de dosage par le disulfonate d'indigo de l'activité de la diamine: oxygéne oxydo-réductase (E.C. 1.4.3.6.) du placenta humain. *Biochimie* **53,** 709,710.
33. McClure, W. R. (1969) A kinetic analysis of coupled enzyme assays. *Biochemistry* **8,** 2782–2786.
34. Storer, A. C. and Cornish-Bowden, A. (1974) The kinetics of coupled enzyme reactions. *Biochem. J.* **141,** 205–209.
35. Corazza, A., Steventano, R., Di Paolo, M. L., Scarpa, M., Mondovi, B., and Rigo, A. (1992) Effect of phosphate ion on the activity of bovine plasma amine oxidase. *Biochem. Biophys. Res. Commun.* **189,** 722–727.
36. Vogel, A. I. (1961) *Quantitative Inorganic Analysis*, Longmans, London.
37. Nelson, D. P. and Kiesow, L. A. (1972) Enthalpy of decomposition of hydrogen peroxide by catalase at 25°C (with molar extinction coefficients of H_2O_2 solutions in the UV). *Anal. Biochem.* **49,** 474–478.
38. Beers, R. F., Jr. and Sizer, I. W. (1952) A spectrophotometric method for the measuring the breakdown of hydrogen peroxide by catalase. *J. Biol. Chem.* **195,** 133–140.

10

Radiochemical Assay of Diamine Oxidase

R. James Storer and Antonio Ferrante

1. Introduction

Diamine oxidase activity (E.C.1.4.3.6) can be measured by a modification of the method first reported by Okuyama and Kobayashi *(1)*, as developed by Tryding et al. *(2–4)*. Oxidation of [1,4-^{14}C]putrescine by diamine oxidase leads to the formation of γ-[^{14}C]-aminobutylaldehyde, which rapidly and spontaneously undergoes an internal cyclization to form the internal aldimine ring compound, Δ^1-[^{14}C]pyrroline **(Fig. 1)**. The pyrroline may undergo further spontaneous polymerization *(5)*. Unlike putrescine, which is very soluble in water, Δ^1-pyrroline and its polymers are soluble in toluene at alkaline pH and may be easily separated from the substrate by solvent extraction. This assay system has been extensively reviewed *(6–10)*.

In biological mixtures γ-aminobutyraldehyde may be alternatively oxidized by aldehyde dehydrogenases (EC 1.2.1.3) to γ-aminobutyric acid (GABA) *(11–13)*. The formation of 4-amino-1-butanol is also possible through reduction by aldehyde dehydrogenase and/or alcohol dehydrogenase *(13,14)*, thus preventing cyclization. Other fates of putrescine in biological mixtures include the acetylation to acetylputrescine by an *N*-acetyltransferase *(11,15,16)* and then oxidation by monoamine oxidase (EC 1.4.3.4) *(11,17)*.

Putrescine may also be metabolized to spermidine by spermine synthase (EC 2.5.1.6) *(11,18,19)*. The Δ^1-pyrroline level decreases in all of the above scenarios and may be interpreted as low DAO activity. Thus, results with this radiochemical method must be regarded with caution if biological mixtures, such as tissue homogenates, are assayed. Blood plasma lacks enzymes that degrade γ-aminobutyraldehyde *(10)*. A number of authors have used enzyme inhibitors to check the validity of the assays *(12,13,18–21)*. However, the use of inhibitors must be judicious, because they may lack specificity and also influence DAO activity *(20)*.

From: *Methods in Molecular Biology, Vol. 79: Polyamine Protocols*
Edited by: D. Morgan Humana Press Inc., Totowa, NJ

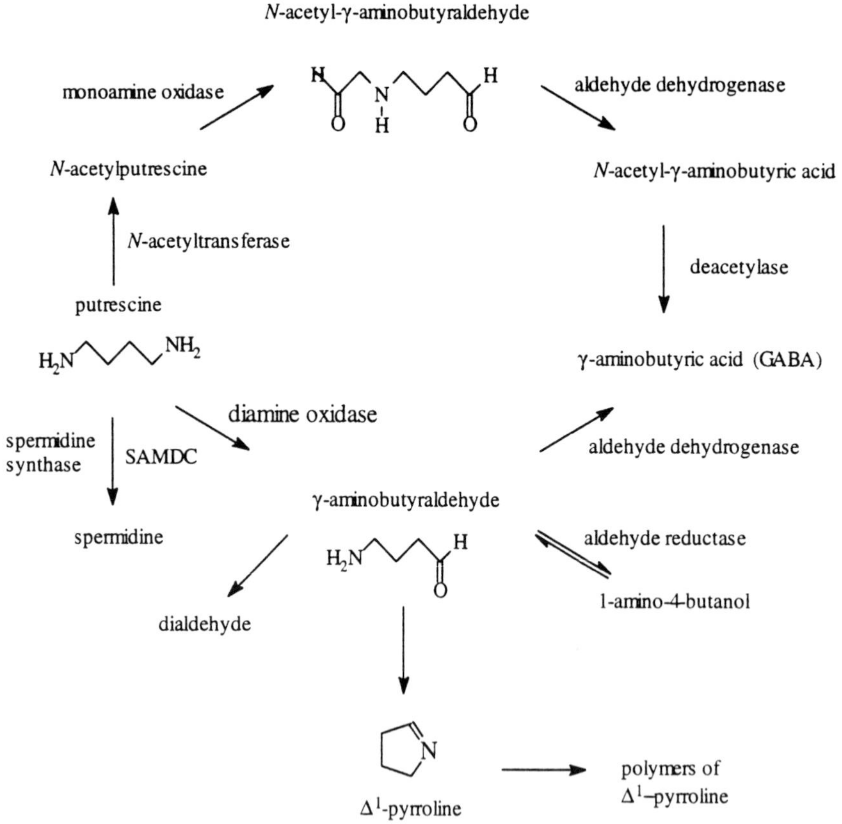

Fig. 1. Fates of putrescine in biological mixtures.

2. Materials

1. Putrescine dihydrochloride (1,4-diaminobutane) is obtained from Sigma (St. Louis, MO).
2. [^{14}C]Putrescine ([1,4-^{14}C]tetramethylenediamine · 2HCl) is obtained from Amersham International (Buckinghamshire, UK).
3. Stock buffer solution: 0.5*M* Tris-HCl, pH 7.4, at 37°C (*see* **Notes 1** and **2**).
4. Putrescine, 10 m*M*: Dissolve 16.1 mg of putrescine dihydrochloride in water and make up to 10 mL.
5. Stock substrate solution: To 250 μL [^{14}C]putrescine add 500 μL of 10 m*M* putrescine to give a total volume of 5.00 mL. The final assay concentration of 102 μ*M* (at 2.45 mCi/mmol), i.e., 0.25 μCi of radioactive putrescine stock and 0.1 μmol cold putrescine in a total assay volume of 1.00 mL.
6. Extraction mixture: toluene containing 1,4-di(2-(5-phenoxazolyl))-benzene (POPOP) at 0.1 g/L and 2,5-diphenoxazole (PPO) at 6.0 g/L (*see* **Note 3**).
7. Screwcap siliconized glass reaction tubes (Pyrex Cat No. 1636; *see* **Note 4**).

8. Miniature polypropylene scintillation vials (Packard Instrument Co., Inc., Meriden, CT).

3. Methods

1. Add 100 µL stock substrate and 100 µL stock buffer to 750 µL water (650 µL water and 100 µL inhibitor solution in the case of inhibition assays) in screwcap siliconized glass reaction tubes.
2. Incubate for 10 min at 37°C for temperature equilibration.
3. Initiate the reactions by adding of 50 µL enzyme sample to triplicate assay tubes.
4. Incubate for 1 h in a 37°C water bath with shaking. Leave the caps off so that the reaction mixtures equilibrate with atmospheric oxygen.
5. Terminate the reactions with 100 µL saturated sodium carbonate solution (*see* **Note 5**).
6. Add 5.0 mL extraction mixture to each tube. Cap the tubes and vortex for 30 s to selectively extract reaction products ***(6)***.
7. Centrifuge the tubes at 400*g* for 5 min using swing out buckets.
8. Freeze the lower, aqueous phase by briefly immersing the tubes in a freezing bath of solid carbon dioxide-alcohol, or liquid nitrogen. Take care not to also freeze the organic phase.
9. Decant the toluene layer into miniature polypropylene vials for scintillation counting in a liquid scintillation spectrometer.

Controls should include a reagent blank and a sample blank in which enzyme activity is terminated at time zero by the addition of 100 µL 72% (w/v) trichloroacetic acid. Ten microliters of substrate solution in scintillant (5 mL) is also counted to determine substrate-specific activity.

4. Notes

1. Use high quality water, such as Milli-Q, for making up buffers.
2. Because the pH of Tris has a significant temperature coefficient it is important to ensure that the buffer will be at the intended pH at the temperature at which it will be used.
3. The extraction mixture is best stored in a dark glass rapid constant volume dispenser.
4. Glass test tubes should be siliconized with proprietary solutions that give water-repellent properties to glassware, such as 2% dimethyldichlorosilane solution in 1,1,1-trichlorothane.
5. This solution may be more effective in stopping the reaction completely if it contains a diamine oxidase inhibitor, such as aminoguanidine (2.5 µ*M* final concentration in the assay). If used, this inhibitor should also be included in the blanks.

References

1. Okuyama, T. and Kobayashi, Y. (1961) Determination of diamine oxidase activity by liquid scintillation counting. *Arch. Biochem. Biophys.* **95,** 242–250.

2. Tryding, N. (1965) Micromethod for determination of plasma diamine oxidase. *Scand. J. Clin. Lab. Invest.* **17(Suppl. 86),** 197.
3. Tryding, N. and Willert, B. (1968) Determination of plasma diamine oxidase (histaminase) in clinical practice. *Scand. J. Clin. Lab. Invest.* **22,** 29–32.
4. Tufvesson, G. and Tryding, N. (1969) Determination of diamine oxidase activity in normal human blood serum. *Scand. J. Clin. Lab. Invest.* **24,** 163–168.
5. Tabor, H. (1951) Diamine oxidase. *J. Biol. Chem.* **188,** 125–136.
6. Smith, C. H., Maghsoudloo, M., Himuna, K., and Murphy, G. M. (1991) Pitfalls in the determination of diamine oxidase activity. *Clin. Chim. Acta.* **201,** 89–94.
7. Neufeld, E. and Chayen, R. (1979) An evaluation of three methods for the measurement of diamine oxidase (DAO) activity in amniotic fluid. *Anal. Biochem.* **96,** 411–418.
8. Beaven, M. A. and Shaff, R. E. (1975) Study of the relationship of histaminase and diamine oxidase activities in various rat tissues and plasma by sensitive isotopic assay procedures. *Biochem. Pharmacol.* **24,** 979–984.
9. Kusche, J., Richter, H., Hesterberg, R., Schmidt, J., and Lorenz, W. (1973) Comparison of the 14C-putrescine assay with the NADH test for the determination of diamine oxidase: description of a standard procedure with a high precision and an improved accuracy. *Agents Actions* **3,** 148–156.
10. Andersson, A.-C., Henningsson, S., Persson, L., and Rosengren, E. (1978) Aspects of diamine oxidase activity and its determination. *Acta Physiol. Scand.* **102,** 159–166.
11. Fogel, W. A. (1986) GABA and polyamine metabolism in peripheral tissues, in *GABAergic Mechanisms in the Mammalian Periphery* (Erdö, S. L. and Bowery, N. G., eds.), Raven, New York, pp. 35–56.
12. Fogel, W. A., Bieganski, T., and Maslinski, C. (1979) Effects of inhibitors of aldehyde metabolizing enzymes on putrescine metabolism in guinea pig liver homogenates. *Agents Actions* **9,** 42–44.
13. Fogel, W. A., Bieganski, T., Wozniak, J., and Maslinski, C. (1978) Interference of aldehyde-metabolising enzymes with diamine oxidase/histaminase/activity as determined by the 14[C] putrescine method. *Biochem. Pharmacol.* **27,** 1159–1162.
14. Dupre, S., Granata, F., Federici, G., and Cavallini, D. (1975) Hydroxyamino compounds produced by the oxidation of diamines with diamine oxidase in the presence of alcohol dehydrogenase. *Physiol. Chem. Phys.* **7,** 517–522.
15. Seiler, N. and Al-Therib, M. J. (1974) Putrescine catabolism in mammalian brain. *Biochem. J.* **144,** 29–35.
16. Seiler, N. and Al-Therib, M. J. (1974) Acetyl-CoA:1,4-diaminobutane *N*-acetyltransferase. Occurrence in vertebrate organs and subcellular localization. *Biochim. Biophys. Acta* **354,** 206–212.
17. Seiler, N. and Eichentopf, B. (1975) 4-aminobutyrate in mammalian putrescine catabolism. *Biochem. J.* **152,** 201–210.
18. Ignesti, G., Perretti, M., and Buffoni, F. (1988) Induction and quail liver diamine oxidase (histaminase). Part I: interference of spermidine synthase on the diamine oxidase activity assay using putrescine as substrate. *Agents Actions* **25,** 37–42.

19. Rao, S. B., Rao, K. S. P., and Mehendale, H. M. (1986) Absence of diamine oxidase activity from rabbit and rat lungs. *Biochem. J.* **234,** 733–736.
20. Sessa, A., Desiderio, M. A., Baizini, M., and Perin, A. (1981) Diamine oxidase activity in regenerating rat liver and in 4-dimethyl-aminoazobenzene-induced and Yoshida AH 130 hepatomas. *Cancer Res.* **41,** 1929–1934.
21. Piacentini, M., Sartori, C., Beninati, S., Barbagli, A. A., and Argento-Ceru, M. P. (1986) Ornithine decarboxylase, transglutaminase, diamine oxidase and total diamines and polyamines in maternal liver and kidney throughout rat pregnancy. *Biochem. J.* **234,** 435–440.

11

Radiochemical Estimation of Polyamine Oxidase

David M. L. Morgan

1. Introduction

Polyamine oxidases, defined in this chapter as amine oxidases capable of catalyzing the oxidation of spermidine or spermine ***(1,2)*** (*see* Chapter 1), are present in plants, bacteria, fungi, protozoa, worms, and all mammalian cells ***(2,3)*** (*see* references in Chapter 1). In mammalian cells spermine and spermidine can be converted back to putrescine by the pathways shown in Chapter 1, **Fig. 5**. The first step is the acetylation of an aminopropyl group to give the N^1-acetyl derivative. This is cleaved by polyamine oxidase to form an aldehyde, 3-acetamidopropanal and either spermidine or putrescine (*see* Chapter 1, **Fig. 6**). However, acetylation is not a prerequisite for oxidation, and nonacetylated spermidine or spermine can also act as substrates, but less efficiently. Oxidation of spermine or spermidine by some tissue enzymes is stimulated if benzaldehyde is added to the assay mixture (**Table 1**). This stimulation results in a lowering of the K_m value for the substrate, probably as a consequence of Schiff base formation between the aldehyde and the primary amino groups of spermine or spermidine, and appears to mimic the effects of in vivo acetylation. Use of benzaldehyde in this assay substitutes for the nonavailability of radiolabeled N^1-acetylspermine or N^1-acetylspermidine, the preferred substrates of tissue polyamine oxidase. The effect of benzaldehyde addition on the particular polyamine oxidase being assayed should always be examined before routinely including it in the assay mixture.

The procedure described here can be used to measure the oxidation of spermine, spermidine, and putrescine (*see also* Chapters 9, 10, and 12), the products being spermidine, putrescine, or a metabolite of putrescine, probably Δ^1-pyrroline, respectively. Putrescine is a diamine and enzymes that oxidize it are often described as diamine oxidases; however, as discussed in Chapter 1,

From: *Methods in Molecular Biology, Vol. 79: Polyamine Protocols*
Edited by: D. Morgan Humana Press Inc., Totowa, NJ

Table 1
The Effect of Benzaldehyde on K_m Values of Rat Liver Polyamine Oxidase for Spermine, and Spermidine *(12–14)*

	Km (μ*M*)	
	–Benzaldehyde	+ Benzaldehyde
Spermidine	50	15
N^1-Acetylspermidine	14	—
Spermine	20	5
N^1-Acetylspermine	0.6	—
N^1,N^{12}-Diacetylspermine	5.0	—

$$NH_2(CH_2)_3NH(CH_2)_4NH(CH_2)_3NH_2 \rightarrow$$

spermine

$$NH_2(CH_2)_3NH(CH_2)_4NH_2 + NH_2(CH_2)_2CHO + H_2O_2$$

spermidine

$$NH_2(CH_2)_3NH(CH_2)_4NH_2 + O_2 \rightarrow NH_2(CH_2)_4NH_2 + NH_2(CH_2)_2CHO + H_2O_2$$

spermidine putrescine

Fig. 1. Oxidation of spermine or spermidine by rat liver polyamine oxidase.

the distinction is not always clear and attempts to separate putrescine-oxidizing activity from spermine- or spermidine-oxidizing activity in, for example, human pregnancy plasma, have not been successful ***(4–6)***.

The assay is based on the oxidation of a radiolabeled polyamine substrate by polyamine oxidase in accordance with Fig. 1. Enzyme and substrate are incubated in an appropriate buffer ***(7,8)*** and then the product(s) are separated from the substrate by ion-exchange chromatography ***(9)*** as shown in Fig. 2. The advantages of this technique are its specificity, in that a defined substrate is used and the accumulation of a specific product is measured. The method can be easily adapted for use with other types of amine oxidase for which the identity of the product is not known, by determining the concentration of HCl necessary to elute the radiolabeled product (**Fig. 3**). The disadvantages are the requirement for a radiolabeled substrate and the limited number of these available. For further discussion and references to other methods *see* Chapters 9 and 10.

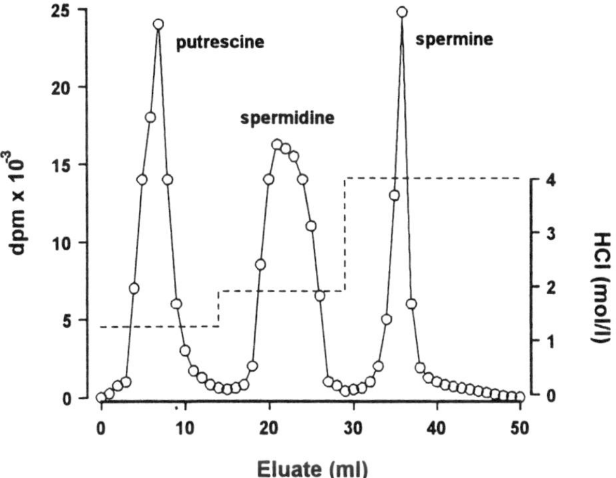

Fig. 2. Separation of [^{14}C]polyamines on Dowex-50 ion exchange resin using a stepwise gradient of HCl. Recovery of radioactivity was >98%.

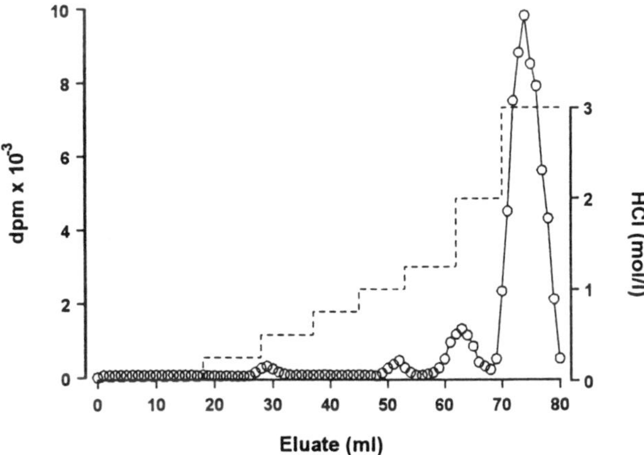

Fig. 3. Elution pattern of [^{14}C]labeled compounds from an assay of polyamine oxidase activity in human pregnancy serum. The major peak corresponds to spermine (unreacted substrate). The two minor peaks to the left comigrate with spermidine and putrescine. The activity eluted with 0.5*M* HCl is a metabolite of putrescine.

2. Materials

1. Pasteur pipets, long form (230 mm).
2. Glass beads, 3-mm diameter.
3. Dowex-50(H^+) ion-exchange resin, 200–400 mesh, 2% crosslinked (Sigma Chemical Co, St. Louis, MO).

4. Spermine tetrahydrochloride (Sigma).
5. Spermidine trihydrochloride (Sigma).
6. Toluene, sulfur-free (Merck Ltd., Poole, UK).
7. Triton X-100 (Merck Ltd.).
8. Benzaldehyde, analytical reagent grade (Merck Ltd.).
9. Triton X-100 (Sigma).
10. Hydrochloric acid, analytical reagent grade (Merck Ltd.).
11. PPO (2,5-diphenyloxazole; Merck Ltd.).
12. POPOP (1,4-bis[2-phenyloxazolyl]-benzene; Merck Ltd.).
13. [^{14}C]spermine (*N,N'*-bis-(3-aminopropyl)-[1,4-^{14}C]tetramethylene-1,4-diamine tetrahydrochloride, 100–124 mCi/mmol, 50 µCi/mL; Amersham International plc Amersham, UK).
14. [^{14}C]spermidine (*N*-(3-[^{3}H]aminopropyl)-[1,4-^{14}C]tetramethylene-1,4-diamine trihydrochloride, 100–124 mCi/mmol, 50 µCi/mL; Amersham).
15. [^{14}C]putrescine ([1,4-^{14}C]tetramethylenediamine dihydrochloride, 100–124 mCi/mmol, 50 µCi/mL; Amersham International plc).
16. Triton-toluene scintillation mixture; add 1400 mL toluene (sulfur-free), with stirring, to 600 mL Triton X-100, then add 10 g PPO and 0.1 g POPOP and stir until dissolved (*see* **Note 1**).
17. 0.5*M* Tris buffer, pH 7.4: Dissolve 119.15 g Tris [Tris(hydroxymethyl)aminomethane] in ≈ 800 mL of distilled water, adjust the pH to 7.4 with HCl, add 10 mL Triton X-100, and make up the volume to 1000 mL.
18. Incubation medium: Add 0.76 mL (7.5 mmol) benzaldehyde to 1000 mL of the 0.5*M* Tris buffer/0.1% Triton X-100 and stir the mixture until homogenous. Store at 4°C, stable for 4 wk.
19. Stock substrate solutions, 10 m*M*: Dissolve 8 mg putrescine dihydrochloride, 12.7 mg spermidine trihydrochloride, or 17.4 mg spermine tetrahydrochloride in 5 mL tissue-culture-grade deionized water, sterilize by filtration through a 0.2-µm filter, and store at 4°C. Shelf life ~6 mo.
20. Buffered substrate solution: Add 50 µL of radiolabeled polyamine (*see* **Note 2**) to 250 µL of the 10 m*M* solution of unlabeled polyamine and make the volume up to 5 mL with incubation medium. The final concentration is 500 µ*M* polyamine with a specific activity ≈1 mCi/m*M*. Prepare fresh for each assay.
21. Trichloroacetic acid, 50% (v/v): Dissolve 50 g trichloroacetic acid in 50 mL of water.
22. Hydrochloric acid: Because the concentration required may vary *(see* **Subheading 3.1.1., step 10***)*, it is convenient to buy standardized 5*M* acid and dilute as necessary.

3. Methods

3.1. Procedure

1. Set aside an aliquot of each sample for protein determination, if required.
2. Preparation of ion-exchange columns:
   a. Fix the required number of 23 cm Pasteur pipet in an upright position (*see* **Note 3**).
   b. Put a 3-mm glass bead into each pipet.
   c. Suspend 10 g of Dowex-50 resin (use as supplied) in 15 mL of water.

   d. Dispense 1 mL of the resin suspension into each column (*see* **Note 4**).
   e. Wash each column with water until the eluate is colorless (about 2.5 mL/column; *see* **Note 5**).
3. Place 200 μL buffered substrate in each assay tube and Incubate at 37°C for 15 min for temperature equilibration.
4. Add 100 μL sample (or 100 μL water in the blanks, used to determine background dpm) and incubate at 37°C for 90 min (*see* **Note 6**).
5. At the end of the incubation period stop the reaction by adding 50 μL of 50% trichloroacetic acid to each tube.
6. Controls, in which the sample is added after the trichloroacetic acid, and blanks, in which the sample is replaced by water, should be included in each assay.
7. Sediment the precipitated protein by centrifugation (>600*g*) for 10 min.
8. Remove a 200-μL aliquot of the reaction mixture supernatant from each assay tube and run it gently onto the top of an ion-exchange column. Allow the liquid level to fall to the top of the resin, but do not let the column run dry. Discard the eluate.
9. Add 200 μL of distilled water to each column, again allow the liquid level to fall to the top of the resin, and discard the eluate.
10. Apply 10 mL of HCl to each column and collect the eluate. The concentration of HCl will depend on the substrate used and should be such that the product(s) are eluted from the ion-exchange resin, but the substrate is retained:

Substrate	Concentration of HCl required to elute product, *M*
Putrescine	0.75
Spermidine	1.25
Spermine	1.9

11. Elute unreacted substrate, if required, using 10 mL of 4.0*M* HCl.
12. Transfer 1 mL of eluate to a polyethylene mini β-vial.
13. Dispense 3 mL of scintillation fluid from a container fitted with a bottle dispenser (e.g., Gilson Distrivar, Anachem Ltd., Luton, UK) into each vial (*see* **Note 7**).
14. Add 10 μL of the buffered substrate solution to three of the β-vials containing 1 mL of HCl of the concentration used to elute the reaction product. These will be used to determine the specific activity of the radiolabeled substrate used.
15. Determine the amount of radioactivity per vial (as disintegrations per minute, dpm) using a liquid-scintillation counter.
16. Dispose of radioactive waste in accordance with local regulations.
17. If the counter is connected to a computer, the data can be saved to disk. Samples, background, and radiolabeled substrate are all counted under the same conditions of quench and efficiency, thus reducing errors and the need for extensive corrections.
18. Perform a protein assay on each sample, if required (for methods *see* refs. ***10,11*** and Chapter 17).

3.2. Calculation of Results

Subtract the background dpm from the dpm of the standards (if this has not already been done by the software in the counter). Divide the resulting value by the number of picomoles of polyamine in the standard (10 μL of a 500 μ*M* solution will contain 5 pmol) to obtain the specific activity in dpm/pmol of polyamine substrate. Subtract the control dpm from the sample dpm and divide the result by the specific activity to obtain picomoles of substrate metabolized during the incubation period; division of this value by the incubation time in minutes will give the enzyme activity in picomoles of substrate oxidized per minute. This can then be related to sample volume or protein content. Because the radiolabel is in the putrescine moiety in [^{14}C]-labeled spermine and spermidine, the labeled metabolites will be retained on the ion-exchange resin during loading and washing (**Subheading 3.**, **steps 8** and **9**); this is not necessarily the case if [^{3}H]-labeled polyamines are used.

4. Notes

1. This is an economical and efficient scintillation fluid but any mixture capable of accepting 25% (v/v) aqueous solution will be satisfactory.
2. The amount of radiolabel should be adjusted in the light of experience to ensure adequate counts in the assay.
3. Plastic-coated (needed for protection from acid spills!) spring clips screwed to a metal bar supported by a lab stand at each end work well.
4. It is important to keep the resin well mixed during this step because it settles out rapidly and the result will be unequal columns.
5. This step will also help to eliminate those columns with a very slow flow rate; not all glass beads are uniform and the bore of the Pasteur pipets will vary, especially between manufacturers! It is advisable to conduct preliminary tests to determine the best sources of beads and pipets and then buy a reasonably large stock of both.
6. The time chosen must be such that the rate of conversion of product to substrate is linear with time and that <2% of the substrate has been consumed
7. Formation of a milky gel may occur but the sample can still be counted successfully.

References

1. Morgan, D. M. L. (1985) Polyamine oxidases. *Biochem. Soc. Trans.* **13,** 322–326.
2. Morgan, D. M. L. (1989) Polyamine oxidase and oxidised polyamines, in *Physiology of the Polyamines*, vol. I (Bachrach, U. and Heimer, Y., eds.), CRC, Boca Raton, FL, pp. 203–229.
3. Müller, S. and Walter, R. D. (1992) Purification and characterization of polyamine oxidase from Ascaris suum. *Biochem J.* **283,** 75–80.
4. Gahl, W. A., Vale, A. M., and Pitot, H. C. (1982) Spermidine oxidase in human pregnancy serum. Probable identity with diamine oxidase. *Biochem. J.* **201,** 161–164.
5. Morgan, D. M. L. (1983) Polyamine oxidation and human pregnancy. *Adv. Polyamine Res.* **4,** 155–167.

6. Morgan, D. M. L. (1985) Human pregnancy-associated polyamine oxidase: partial purification and properties. *Biochem. Soc. Trans.* **13,** 351,352.
7. Morgan, D. M. L. and Ilei, G. (1981) Radiochemical estimation of serum polyamine oxidase activity in human pregnancy. *Med. Lab. Sci.* **38,** 49–56.
8. Ferrante, A., Storer, R. J., and Cleland, L. J. (1990) Polyamine oxidase activity in rheumatoid arthritis synovial fluid. *Clin. Exper. Immunol.* **80,** 373–375.
9. Inoue, H. and Mizutani, A. (1973) A new method for isolation of polyamines from animal tissues. *Anal. Biochem.* **56,** 408–416.
10. Lowry, O. H., Rosenbrough, N. J., Farr, A. L., and Randall, R. J. (1951) Protein measurement with the Folin phenol reagent. *J. Biol. Chem.* **193,** 265–275.
11. Bradford, M. M. (1976) A rapid and sensitive method for the quantitation of microgram quantities of protein utilizing the principle of protein-dye binding. *Anal. Biochem.* **72,** 248–254.
12. Höltta, E. (1983) Polyamine oxidase (rat liver). *Methods Enzymol.* **94,** 306–311.
13. Höltta E. (1977) Oxidation of spermidine and spermine in rat liver: purification and properties of polyamine oxidase. *Biochemistry* **16,** 91–100.
14. Bolkenius, F. N. and Seiler, N. (1981) Acetylderivatives as intermediates in polyamine catabolism. *Int. J. Biochem.* **13,** 287–292.

12

Measurement of Aldehyde Formation as an Assay for Polyamine Oxidase Activity

David M. L. Morgan

1. Introduction

The polyamines spermine and spermidine can be metabolized by polyamine oxidases in two ways, oxidative deamination or oxidative cleavage ***(1,2)*** (*see* Chapter 1). An example of the former is the amine oxidase present in bovine (and other ruminant) sera ***(3)*** that deaminates polyamines in accordance with **Fig. 1**. In contrast, tissue polyamine oxidase (typified by rat liver polyamine oxidase) ***(4)*** cleaves spermine or spermidine with the release of an aminopropyl moiety (**Fig. 2**).

The enzyme-containing sample is incubated with substrate, the aldehydes formed are trapped by reaction with NBTH (*N*-methyl-2-benzothiazolone hydrazone hydrochloride), and the resulting derivative is converted to a blue-colored complex by the addition of ferric ions ***(5,6)***.

Disadvantages of this method are:

1. Oxidation of polyamines by some plant enzymes may not result in aldehyde formation ***(8,9)*** and other assays, such as hydrogen peroxide measurement (*see* Chapter 9), must be used.
2. In some micro-organisms such as *Serratia marcescens*, spermidine may be metabolized also by dehydrogenative cleavage ***(10)***.
3. Aminopropionaldehyde (3-aminopropanal), if formed, is a substrate for aminoaldehyde dehydrogenase ***(11)*** and the presence of this enzyme in the sample can lead to erroneously low results.
4. The sensitivity is less than the peroxide/fluorescence method (*see* Chapter 9).

The advantages of the method, which was originally devised to measure polyamines ***(6)***, are that it is simple, it requires no special apparatus, it is easily adapted for use with small volumes in 96-well microplates, it can be used to

From: *Methods in Molecular Biology, Vol. 79: Polyamine Protocols*
Edited by: D. Morgan Humana Press Inc., Totowa, NJ

$NH_2(CH_2)_3NH(CH_2)_4NH(CH_2)_3NH_2 + 2O_2 + 2H_2O \rightarrow$
spermine
$OHC(CH_2)_2NH(CH_2)_4NH(CH_2)_2CHO + 2NH_3 + 2H_2O_2$

$NH_2(CH_2)_4NH(CH_2)_3NH_2 + O_2 + H_2O \rightarrow NH_2(CH_2)_4NH(CH_2)_2CHO +$
spermidine
$NH_3 + H_2O_2$

Fig. 1. Oxidation of spermine or spermidine by bovine serum amine oxidase.

$NH_2(CH_2)_3NH(CH_2)_4NH(CH_2)_3NH_2 + O_2 + H_2O \rightarrow$
spermine
$NH_2(CH_2)_3NH(CH_2)_4NH_2 + NH_2(CH_2)_2CHO + H_2O_2$
spermidine

$NH_2(CH_2)_4NH(CH_2)_3NH_2 + O_2 + H_2O \rightarrow NH_2(CH_2)_4NH_2 +$
spermidine
$NH_2(CH_2)_2CHO + H_2O_2$
putrescine

Fig. 2. Oxidation of spermine or spermidine by rat liver polyamine oxidase.

screen rapidly large numbers of samples (e.g., from enzyme purification or other forms of column chromatography, *see* **ref. 7**) and the substrates spermine or spermidine are relatively inexpensive.

2. Materials

1. NBTH (3-methyl-2-benzothiazolone hydrazone hydrochloride hydrate; Aldrich Chemical Co. [Poole, UK]): Dissolve 0.4 g in 100 mL distilled water. Store at 4°C and use within 1 wk; discard sooner if discolored.
2. 0.2*M* Tris-HCl buffer, pH 7.4 (*see* **Note 1**).
3. Stock substrate solutions, 10 m*M*: Dissolve 12.7 mg spermidine trihydrochloride or 17.4 mg spermine tetrahydrochloride in 5 mL deionized water and store at 4°C. Shelf life ~6 mo.
4. Buffered substrate solution: Add 0.5 mL of 10 m*M* polyamine solution to 9.5 mL of the Tris buffer; the concentration of polyamine will be 500 µ*M*.
5. 0.2% ferric chloride solution: Dissolve 0.2 g ferric chloride in 0.01*M* HCl and make up to 100 mL. Any turbidity should be removed by filtration through a Whatman No. 3 paper or equivalent. Discard if turbidity because of formation of hydroxides returns.

3. Methods

1. Set aside an aliquot of each sample for protein determination, if required.
2. Place 100 µL buffered substrate in each assay tube, add 100 µL of sample, and incubate at 37°C for 60 min (*see* **Note 2**).

3. Blanks, in which the sample is replaced by sample buffer or water, should be included in each assay.
4. Add 2.5 mL of 0.2% ferric chloride to each tube and stand for a further 15 min (*see* **Note 3**).
5. Measure the intensity of the color at 660 nm against a blank.

3.1. Calculation of Results

Enzyme activities are usually reported as quantity of substrate converted in unit time and related to quantity of enzyme protein or volume of sample.

$$c = (\Delta E \times V)/(\varepsilon \times v) \quad (1)$$

where c = the concentration of aldehyde in the assay in mol/L, ΔE is the difference in absorbance between the sample and the blank; V is the volume of the sample (in mL); ε is the molar absorbtivity (i.e., the absorbance of a 1 mol/L solution measured using a 1 cm light path), and v is the volume of sample used in the assay. The molar absorbance of the colored complex obtained after the oxidation of spermine is 12.5×10^3 L/mol/cm, and for spermidine the value is 6.25×10^3, based on the formation of a dialdehyde from spermine and a monoaldehyde from spermidine (Fig. 1). The calculated concentration of aldehyde must then be divided by the incubation time in minutes to give moles of product formed per minute. This value can then be related to either the volume or protein content of the sample.

4. Notes

1. The pH of the incubation buffer should be adjusted to the optimum for the enzyme being assayed (if known!).
2. The time chosen should be such that the rate of conversion of product to substrate is linear with time and that <2% of the substrate has been consumed.
3. The volumes are sufficient for use with 1 cm cuvets; if smaller cuvets are used, the volume of ferric chloride can be reduced with a consequent increase in sensitivity.

References

1. Morgan, D. M. L. (1985) Polyamine oxidases. *Biochem. Soc. Trans.* **13,** 322–326.
2. Morgan, D. M. L. (1989) Polyamine oxidase and oxidised polyamines, in *Physiology of the Polyamines*, vol. I (Bachrach, U. and Heimer, Y., eds.), CRC, Boca Raton, FL, pp. 203–229.
3. Tabor, C. W., Tabor, H., and Bachrach, U. (1964) Identification of the aminoaldehydes produced by the oxidation of spermine and spermidine with purified plasma amine oxidase. *J. Biol. Chem.* **239,** 2194–2203.
4. Höltta, E. (1977) Oxidation of spermidine and spermine in rat liver: purification and properties of polyamine oxidase. *Biochemistry* **16,** 91–100.

5. Sawicki, E., Hauser, T. R., Stanley, T. W., and Elbert, W. (1961) The 3-methyl-2-benzothiazolone hydrazone test. Sensitive new methods for the detection, rapid estimation, and determination of aliphatic aldehydes. *Anal. Chem.* **33,** 93–96.
6. Bachrach, U. and Reches, B. (1966) Enzymic assay for spermine and spermidine. *Anal. Biochem.* **17,** 38–48.
7. Morgan, D. M. L., Bachrach, U., Assaraf, Y. G., Harari, E., and Golenser, J. (1986) The effect of purified aminoaldehydes produced by polyamine oxidation on the development *in vitro* of *Plasmodium falciparum* in normal and glucose-6-phosphate-dehydrogenase-deficient erythrocytes. *Biochem. J.* **236,** 97–101.
8. Smith, T. A. (1985) The di- and polyamine oxidases of higher plants. *Biochem. Soc. Trans.* **13,** 319–322.
9. Rinaldi, A., Floris, G., and Giartosio, A. (1985) Plant amine oxidases, in *Structure and Functions of Amine Oxidases* (Mondovi, B., ed.), CRC, Boca Raton, FL, pp. 51–62.
10. Tabor, C. W. and Kellogg, P. D. (1970) Identification of flavin adenine dinucleotide and heme in a homogenous spermidine dehydrogenase from Serratia marcescens. *J. Biol. Chem.* **245,** 5424–5433.
11. Padmanabhan, R. and Tchen, T. T. (1972) Aminoaldehyde dehydrogenase from a *Pseudomonas* species grown on polyamines. *Arch. Biochem. Biophys.* **150,** 531–541.

IV

Measurement of Polyamines

13

Determination of Polyamines as Their Benzoylated Derivatives by HPLC

David M. L. Morgan

1. Introduction

Almost all of the methods available to analytical biochemists have been applied to polyamine analysis at some time. The diversity of the methods used is an indication of the difficulties experienced in attempts to measure the polyamine content of biological samples. The quantitation of polyamines by high-pressure liquid chromatography (HPLC) of their benzoyl derivatives was described by Redmond and Tseng *(1)* in 1979 and has been modified at intervals ever since *(2–12)*. The advantages of this technique are its relative simplicity and the need for only very basic HPLC equipment—pump, column, injection valve, and UV-detector. The main disadvantage, as noted by Seiler *(13,14)*, is that benzoyl chloride, in addition to undergoing condensation with primary and secondary amines (**Fig. 1**), may also react with imidazole nitrogens and phenolic hydroxyl groups. However, benzoylation results in fewer side reactions than are seen with dansyl chloride and hence extensive clean-up procedures may not be necessary. The main contaminants in the chromatogram are benzoyl chloride, benzoic acid, benzoic anhydride, and methylbenzoate, the first two being quantitatively the most important. Contrary to the accepted dogma, some benzoyl chloride may survive the derivatization procedure and be extracted and applied to the column; its presence can be detected by a peak in the chromatogram (**Fig. 2**) and by its odor if the top of the column is unscrewed; use of a guard column (or cartridge) will reduce the deleterious effects benzoyl chloride on the analytical column. Benzoic anhydride may comigrate with spermine, benzoic acid can interfere with the separation of *N*-acetylputrescine and *N*-(3-aminopropyl)pyrrolidin-2-one *(7)*, and methyl benzoate can overlap N^1-acetylspermidine, depending on the solvent mixture used.

From: *Methods in Molecular Biology, Vol. 79: Polyamine Protocols*
Edited by: D. Morgan Humana Press Inc., Totowa, NJ

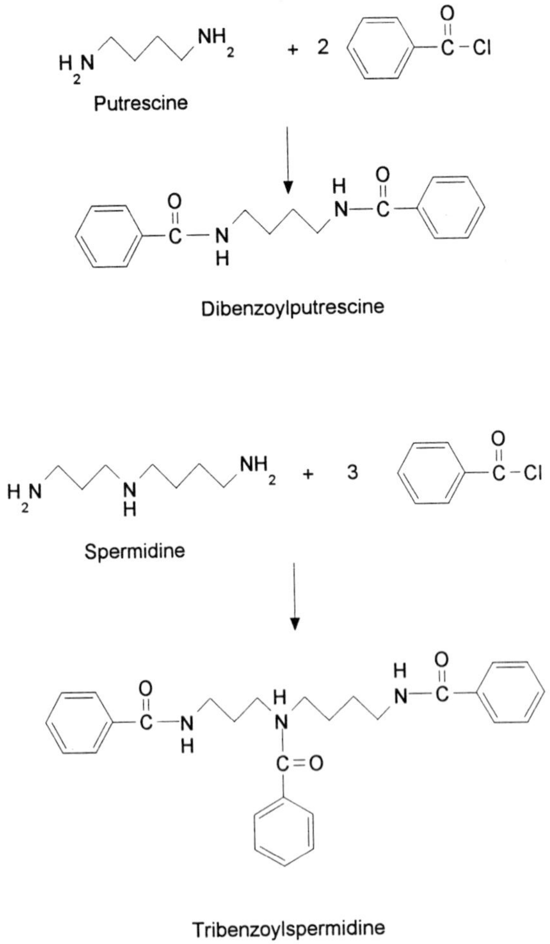

Putrescine

Dibenzoylputrescine

Spermidine

Tribenzoylspermidine

Fig. 1. The reaction between benzoyl chloride and putrescine or spermidine under alkaline conditions (Schotten-Baumann condensation).

Clean-up steps have been recommended by some authors ***(12,15)*** and do result in the reduction/removal of some contaminating peaks. However, if these do not interfere then this step can be omitted. Methanol-water is the most widely used solvent system and most reported separations have been performed isocratically ***(1–5,7,8,10–12)***, obviating the need for a gradient maker or two pump system, although use of a gradient may enable the simultaneous detection of other compounds ***(6,9)***. Acetonitrile-water mixtures, also used isocratically, can improve sensitivity ***(11)***. The characteristics of the assay have been extensively investigated ***(4,7,12)***.

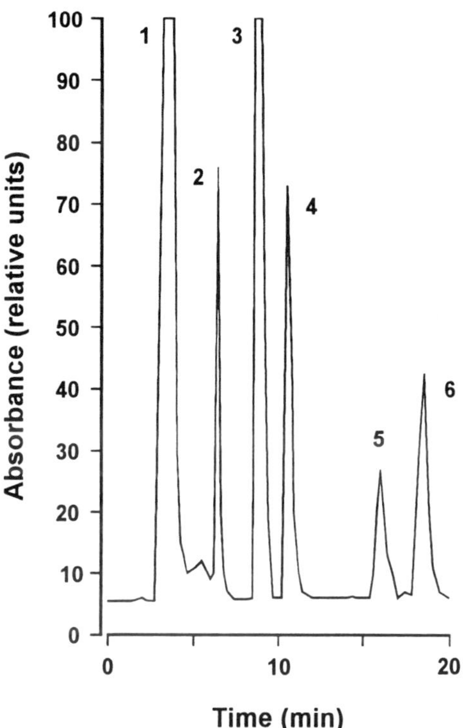

Fig. 2. Chromatogram of benzoylated polyamines. The peaks, which were identified by individual analyses, are 1, benzoic acid; 2, putrescine; 3, benzoyl chloride; 4, spermidine; 5, diaminooctane; 6, spermine. Flow rate 0.4 mL/min, detection at 229 nm, chart speed 300 mm/h, full scale deflection = 0.1 absorbance.

The procedure described here has been used to follow intracellular changes in the concentrations of putrescine, spermidine, and spermine in endothelial cells *(16)* and macrophages in response to exogenous polyamines.

2. Materials

1. HPLC pump, Waters model 510 (Millipore UK Ltd., Watford, UK) or equivalent.
2. Rheodyne syringe injection port, Model 7125, fitted with 100 µL sample loop (Anachem Ltd., Luton, UK).
3. HPLC column: Spherisorb ODS2 5 µ, 250 × 4.6-mm, fitted with a 50 × 4.6 mm guard cartridge (Anachem Ltd.; *see* **Note 1**).
4. Waters Programmable detector M490, or equivalent.
5. Strip-chart recorder or datalogger and integrator.
6. Injection syringe for use with Rheodyne valves, 50 µL (Anachem Ltd.).
7. 1,8-Diaminooctane (*see* **Note 2**).

8. Benzoyl chloride, analytical reagent grade.
9. Methanol, HPLC grade.
10. Water, HPLC grade.
11. 2*M* Sodium hydroxide solution.
12. Trichloroacetic acid: 50% (v/v); add 50 g trichloroacetic acid to 50 mL water and stir until dissolved.
13. 1 m*M* polyamine standards: Dissolve 8 mg putrescine dihydrochloride, 12.7 mg spermidine trihydrochloride, or 17.4 mg spermine tetrahydrochloride in 50 mL distilled/deionized water and store at 4°C.
14. Internal standard: Dissolve 21.5 mg of 1,8-diaminooctane dihydrochloride in water and dilute to 100 mL; final concentration, 1 m*M*. Store at 4°C.
15. Polyamine working standard: Add 1 mL of each polyamine solution to 1 mL of the internal standard solution and dilute to 10 mL.
16. HPLC mobile phase: Add 600 mL of methanol (HPLC grade) to 400 mL of HPLC grade water and mix. Place the reservoir containing the solvent mixture in an ultrasonic bath and degas by sonnication for 5 min (*see* **Note 3**).

3. Methods

3.1. Preparation of Cell Extract

1. Adherent cells:
   a. Remove the culture medium.
   b. Rinse the cell layer twice with PBS.
   c. Add 1 mL PBS (*see* **Note 4**) and detach the cell by scraping with a cell scraper (e.g., Costar 3010, Cambridge, MA).
   d. Transfer the cell suspension to a 1.5-mL Eppendorf tube.
2. Cells grown in suspension:
   a. Separate the cells from the culture medium by low-speed centrifugation (~150*g* for 5 min).
   b. Resuspend the cells twice in PBS and sediment by centrifugation.
   c. Resuspend the cells in 1 mL PBS (*see* **Note 4**).
3. Add 100 µL of 50% TCA and 50 µL (50 nmol) of the internal standard to each tube.
4. Thoroughly mix on a vortex shaker.
5. Sediment the precipitated protein and cell debris by centrifugation in an Eppendorf centrifuge (12,000*g*, 1 min).
6. Remove the supernatant and store at –20°C until required.

3.2. Benzoylation Procedure

1. Place 500 µL of TCA cell extract or polyamine standard, in a 16 × 100 mm glass tube with a PTFE-lined screw cap (e.g., Corning 9826-16).
2. Add 2 mL of 2*M* NaOH and mix.
3. Add 10 µL of 50% benzoyl chloride in methanol (≈40 µmol; *see* **Note 5**).
4. Mix vigorously to disperse the benzoyl chloride (1–2 min on a vortex stirrer).
5. Allow the reaction to proceed at room temperature for 30 min (timing is not

critical, can be up to 2 h, but must be standardized) with either continuous or intermittent shaking (e.g., place the tube rack on an oscillating shaker or vortex for 1 min at 5 min intervals).

6. Add 1 mL of chloroform to each tube and vortex for 1 min.
7. Centrifuge briefly to separate the layers (or leave to stand, but this may take some time, depending on the amount of cellular material present).
8. Transfer the lower, chloroform layer to a clean glass tube.
9. Repeat **steps 6–8** and combine the chloroform extracts.
10. Add 1 mL of HPLC grade water to the chloroform extracts.
11. Mix thoroughly (vortex for 1 min).
12. Centrifuge briefly to separate the layers.
13. Transfer as much as possible of the lower, chloroform layer to a clean glass tube but do not include any of the aqueous layer (*see* **Note 6**).
14. Evaporate the chloroform extracts to dryness. This should be done in a fume cupboard, either using a stream of nitrogen or air (*see* **Note 7**).
15. Dissolve the dry residue of benzoylated polyamines in 100 µL of mobile phase (60% methanol–40% water) and store at –20°C.

3.3. Calibration of HPLC Column

1. Derivatize separately 50 µL (50 nmol) of each polyamine standard and diaminooctane (the internal standard) plus 450 µL water according to **Subheading 3.2., steps 1–15**.
2. Derivatize 500 µL of the working standard in the same way.
3. Set the detector to 229 nm (*see* **Note 8**).
4. Set the recorder so that 0.1 absorbance units gives full-scale deflection.
5. Set the flow rate to 0.4 mL/min.
6. Inject 5, 10, and 25 µL (≅2.5, 5, and 10 nmol) of each and determine that the peak heights/areas increase proportionally and that the separation of peaks in the mixed standard is satisfactory. Adjust detector and recorder sensitivity and sample volume as necessary.
7. Run each chromatogram for 25 min and equilibrate the column by pumping mobile phase for a further 10 min after the emergence of the last peak.
8. Note the retention times of each peak as an aid to subsequent identification (*see* **Note 9**).

3.4. HPLC Analysis

1. Start each analysis by injecting derivatized working standard.
2. Check that the resulting chromatogram is satisfactory.
3. Inject 25–50 µL of the first sample and observe the result.
4. Adjust sample volume and recorder/detector sensitivity if necessary.
5. Repeat injection of sample 1.
6. Continue to assay samples.
7. Run the standard mixture after every 5–10 samples.
8. After the last assay, pump mobile phase through the column overnight.

3.5. Calculation of Results

Measure the area of each peak (*see* **Note 10**). If 10 μL of the working standard was used, then each peak corresponds to an initial quantity of 5 nmol of polyamine. Any losses arising from sample processing can be compensated by use of the internal standard (diaminooctane, DAO), e.g., putrescine:

$$\text{Amount of putrescine in volume of sample applied to column} = \frac{(\text{peak area of putrescine in sample} \times \text{peak area of DAO in sample})}{(\text{peak area of putrescine in standard} \times \text{peak area of DAO in standard})} \quad (1)$$

From a knowledge of the amount of sample represented by the volume applied to the column, the concentration of polyamine (e.g., putrescine) in the sample can be calculated.

4. Notes

1. A new column should be equilibrated by pumping mobile phase through it overnight.
2. Purification of 1,8-diaminooctane: Dissolve the commercial product in the minimum amount of water, neutralize the resulting solution using 5*M* HCl, then slowly add ethanol to the stirred solution until a white precipitate forms. Collect the precipitate by filtration, wash with ethanol, and allow to dry in air at room temperature. Store over desiccant at 4°C. The 1,8-diaminooctane dihydrochloride is stable for at least 1 yr.
3. Sonication is a cheap and effective alternative to degassing with helium ***(10)***. Failure to degas the mobile phase will result in spurious peaks throughout the chromatogram. Inclusion of a 0.2-μm filter (Solvent In-Line Filter/Degasser, Whatman International Ltd, Maidstone, UK) in the solvent line from reservoir to pump is also recommended, both to maintain the degassed state of the mobile phase and to prolong column life.
4. The volume of PBS should be adjusted in proportion to cell number; 1 mL is adequate for a 75 cm^2 tissue culture flask, 500 μL for a 25 cm^2 flask, and so on.
5. This reagent should be handled with care; it is a lacrymator, irritating to skin, eyes, and mucous membranes. Ideally, benzoyl chloride should be present in ~10% molar excess of the amount required to react with the polyamines in the sample; use of too much can cause difficulties *(see* **Note 6***)* and the quantity used should be adjusted in the light of experience. It can be difficult to pipet small volumes of benzoyl chloride; use of a 50% (v/v) solution in methanol will ease dissolution of the reagent in the reaction mixture and enable small adjustments to be made in the amount used. The solution should be prepared fresh on the day.
6. It is important not to include any of the aqueous layer in the transfer because it may result in spurious peaks and residual benzoyl chloride can react with and damage the stationary phase. Incomplete transfer of the chloroform layer will not negate the assay; the peak heights will be lowered but the presence of the internal standard will allow compensation for this in the calculation of results.

7. In the author's laboratory this is done by means of a series of long-form (23 cm) Pasteur pipet attached to a manifold that is connected to a filter pump. The manifold is adjusted so that the tips of the pipets are close to the surface of the chloroform and the pump is used to draw room air into the tubes and sweep out the chloroform vapor.
8. Choice of wavelength for detection is a compromise between sensitivity and solvent absorbance. Methanol has an absorbance of 1.000 at 205 nm, decreasing to 0.160 at 225 nm, 0.020 at 250 nm, and 0.005 at 300 nm ***(17)***; whereas the absorbance of benzoyl putrescine is maximal at 229 nm ***(4)***.
9. Retention times, which varied with the age of the column, were: putrescine 6.75–7.0 min, spermidine 11.0–11.25, and spermine 19.0–20.25; diaminooctane emerged at 17.5–18.3 min, conveniently between spermidine and spermine. Benzoic acid eluted at 3.0–3.25 min and benzoyl chloride at 9.25–9.8 min (**Fig. 2**).
10. In the absence of an integrator, a surprisingly accurate assessment of peak areas can be made by using scissors to carefully cut out each peak from the chart and then weighing the cutout paper on a sensitive balance.

References

1. Redmond, J. W. and Tseng, A. (1979) High-pressure liquid chromatographic determination of putrescine, cadaverine, spermidine and spermine. *J. Chromatogr.* **170,** 479–481.
2. Flores, H. E. and Galston, A. W. (1982) Analysis of polyamines in higher plants by high performance liquid chromatography. *Plant Physiol.* **69,** 701–706.
3. Clarke, J. R. and Tyms, A. S. (1986) Rapid analysis of polyamines in cell culture by high performance liquid chromatography. *Med. Lab. Sci.* **43,** 258–261.
4. Verkolen, C. F., Romijin, J. C., Schroeder, F. H., van Schalkwijk, W. P., and Splinter, T. A. W. (1988) Quantitation of polyamines in cultured cells and tissue homogenates by reversed-phase high-performance liquid chromatography of their benzoyl derivatives. *J. Chromatogr.* **426,** 41–54.
5. Wongyai, S., Oefner, P., and Bonn, G. (1988) HPLC analysis of polyamines and their acetylated derivatives in the picomole range using benzoyl chloride and 3,5-dinitrobenzoyl chloride as derivatizing agent. *Biomed. Chromatogr.* **2,** 254–257.
6. Slocum, R. D., Flores, H. E., Galston, A. W., and Weinstein, L. H. (1989) Improved method for HPLC analysis of polyamines, agmatine and aromatic monoamines in plant tissue. *Plant Physiol.* **89,** 512–517.
7. Wongyai, S., Oefner, P., and Bonn, G. (1989) High resolution analysis of polyamines and their acetyl derivatives using HPLC. *J. Liquid Chromatogr.* **12,** 2249–2261.
8. Watanabe, S., Saito, T., Sato, S., Nagase, S., Ueda, S., and Tomita, M. (1990) Investigation of interfering products in the high-performance liquid chromatographic determination of polyamines as benzoyl derivatives. *J. Chromatogr.* **518,** 264–267.
9. Kotzabasis, K., Christakis-Hampsas, M. D., and Roubelakis-Angelakis, K. A. (1993) A narrow-bore HPLC method for the identification and quantitation of free, conjugated, and bound polyamines. *Anal. Biochem.* **214,** 484–489.

10. Taibi, G. and Schiavo, M. R. (1993) Simple high-performance liquid chromatographic assay for polyamines and their monoacetyl derivatives. *J. Chromatogr.* **614,** 153–158.
11. Mei, Y.-H. (1994) A sensitive and fast method for the determination of polyamines in biological samples. Benzoyl chloride pre-column derivatization high-performance liquid chromatography. *J. Liquid Chromatogr.* **17,** 2413–2418.
12. Schenkel, E., Berlaimont, V., Dubois, J., Helson-Cambier, M., and Hanocq, M. (1995) Improved high-performance liquid chromatographic method for the determination of polyamines as their benzoylated derivatives: application to P388 cancer cells. *J. Chromatogr. B. Biomed. Appl.* **668,** 189–197.
13. Seiler, N. (1983) Liquid chromatographic methods for assaying polyamines using prechromatographic derivatization. *Methods Enzymol.* **94,** 10–25.
14. Seiler, N. (1986) Polyamines. *J. Chromatogr.* **379,** 157–176.
15. Wehr, J. B. (1995) Purification of plant polyamines with anion-exchange column clean-up prior to high-performance liquid chromatographic analysis. *J. Chromatogr.* **709,** 241–247.
16. Bogle, R. G., Mann, G. E., Pearson, J. D., and Morgan, D. M. L. (1994) Endothelial polyamine uptake: selective stimulation by L-arginine deprivation or polyamine depletion. *Am. J. Physiol.* **266,** C776–C783.
17. Przybytek, J. T. (1982) *High Purity Solvent Guide.* Burdick and Jackson Laboratories Inc., Muskegon, MI.

14

A Dansyl Chloride-HPLC Method for the Determination of Polyamines

Karl J. Hunter

1. Introduction

The use of the fluorescent acid chloride 5-dimethylaminonaphthalene-1-sulfonylchloride (dansyl chloride) is particularly suited to the separation and quantitation of polyamines by chromatographic techniques. Both the primary and secondary amino groups of polyamines are derivatized, giving a high fluorescence yield, and the derivatization reaction is quick and simple to perform. Using this method, polyamines can be detected and quantitated in many biological matrices, including cerebrospinal fluid, urine, blood, and tissue lysates *(1,2)*; for CSF and urine this method can be used clinically *(1)*. When combined with an acid hydrolysis step *(1,3)*, reproducible results can be obtained for both free and conjugated polyamines *(1,4)*.

2. Materials

1. Water-saturated ethyl acetate: Shake ethyl acetate and deionized water together, allow the mixture to settle, and use the top organic phase.
2. Propionic/hydrochloric acid: Propionic acid and 12*M* constant boiling hydrochloric acid (Pierce Ltd., UK) are carefully mixed in equal volumes. The mixture is stable at room temperature, but should be made freshly for each analysis.
3. Dansyl chloride: This is dissolved in "Analar" acetone at 10 mg/mL and stored in a foil-wrapped container, because the compound is light-sensitive. This solution is stable over the course of a day at room temperature. The dry powder is stored at –20°C, but eventually becomes difficult to dissolve fully in acetone because of degradation and should be replaced.
4. Proline solution: Proline is dissolved in water as a 25% (w/v) solution. This can be stored at –20°C for several months.

From: *Methods in Molecular Biology, Vol. 79: Polyamine Protocols*
Edited by: D. Morgan Humana Press Inc., Totowa, NJ

5. The two high-pressure liquid chromatography (HPLC) solvents are solvent A: 10 m*M* sodium dihydrogen phosphate; solvent B: HPLC grade acetonitrile.
6. DAH Internal standard: 1,7-Diaminoheptane (DAH) is converted to the nonhygroscopic dihydrochloride by dissolving in an excess of 1*M* hydrochloric acid, followed by lyophilization. The dihydrochloride is then dissolved in 10 m*M* HCl at a concentration of 400 μ*M*. This solution can be stored at –20°C for several months.
7. Standard polyamine solution: Putrescine dihydrochloride, DAH, spermidine trihydrochloride, and spermine tetrahydrochloride are dissolved in 10 m*M* HCl at 0.8 m*M*. Any other required polyamines are made up at the same concentration. This solution can be stored at –20°C for several months.
8. An HPLC system equipped with a 5 μm ODS reversed-phase column (a 4.6 × 300 mm ion-paired column from Beckman Instruments Ltd. [High Wycombe, UK] works well), preferably with a guard column (e.g., Newguard RP-18 guard column from Anachem Ltd.) coupled to an in-line fluorescence detector equipped with a microvolume flow cell.

3. Methods

1. Spin down the cells to be analyzed in a 1.5-mL screw-capped microcentrifuge tube (Greiner Labortechnik Ltd., Dursley, UK) and wash 2–3 times in an appropriate buffer; e.g., phosphate-buffered saline for mammalian cells, or a salt solution, 0.9% (w/v) sodium chloride for bacteria.
2. Add 250 μL of water and resuspend the pellet, then add 250 μL 10% trichloroacetic acid dissolved in 0.01*M* HCl and mix. (The hydrochloric acid prevents polyamines from binding to glass as the pH increases). Add 10 μL of the DAH internal standard (400 μ*M* DAH). Leave on ice for 30 min to allow the protein to precipitate. Centrifuge and remove the supernatant. Extract the supernatant five times with an equal volume of water-saturated ethyl acetate to remove excess trichloroacetic acid. If the free polyamine content only is required, the solution is then lyophilized in the microfuge tube. The residue can be redissolved in 50 μL 10 m*M* HCl with sonication if necessary.
3. If the total polyamine content (free and conjugated) is required, after extraction with ethyl acetate the solution is transferred to a reactivial (Pierce Ltd.) and lyophilized. Add 50 μL propionic/hydrochloric acid to the reactivial and heat at 160°C in a heating block for 15 min. Allow to cool and lyophilize. Redissolve in 50 μL 10 m*M* HCl with sonication.
4. To the 50 μL redissolved extract in a 1.5 mL screw-capped centrifuge tube, add 200 μL saturated sodium carbonate solution and 200 μL of the dansyl chloride solution. (If the sample has been hydrolyzed, the lysate can be transferred from the reactivial to a centrifuge tube and the reactivial washed out with the sodium carbonate solution to ensure complete transfer.)
5. Mix for 10 s and heat at 65–70°C for 10 min. After allowing to cool, add 100 μL of 25% (w/v) proline solution and mix thoroughly. Microcentrifuge for 2 min and remove the top acetone layer, which is clear or pale yellow. The volume is usually 350 μL, but can be measured using a Gilson pipet. The bottom layer is

often solid, but is nonfluorescent and does not contain any of the dansyl derivatives. Transfer the top acetone layer to another tube. This solution can also be stored in the dark at room temperature, but the solutions tend to degrade over the course of the day.

6. Inject 20 µL onto the HPLC. If it is necessary to dilute the sample, acetone can be used, but this tends to force water out of solution, altering the concentration of polyamines. Therefore, the standard should be similarly treated.

The running conditions are as follows:

1. Temperature: ambient
2. Elution gradient:

on injection:	45% B to 80% B over 14 min.
at 14 min:	80% B to 90% B over 1 min.
at 15 min:	hold at 90% B for 7 min
at 22 min:	90% B to 45% B over 1 min. (Stop data).
at 29 min:	Next sample injected.

3. Detection: fluorescence; 340 nm (excitation), 515 nm (emission).
4. Calibration: A solution of 50 µL 10 m*M* HCl is dansylated as a blank. This is needed because the method produces spurious peaks as well as a large reagent peak eluting at the solvent front. A standard is also dansylated. This consists of 4 nmol of each polyamine to be assayed (usually putrescine, DAH, spermidine, and spermine) in 50 µL 10 m*M* HCl. An injection volume of 20 µL will then give decent peaks.

The peak area of the polyamines in the sample can be standardized for experimental loss using the DAH peak, as shown below. Given that the polyamine peaks in the standard were produced by an initial solution of 4 nmol polyamine in 50 µL (80 µ*M*), the concentration of each polyamine in the sample can be ascertained from its peak area in the sample and standard chromatogram, i.e.,

$$\text{Concentration (nmol/50 µL)} = (\text{peak area in sample} \times \text{peak area of DAH in sample} \times 4)/(\text{peak area in standard} \times \text{peak area of DAH in standard}) \quad (1)$$

Knowing the number of cells or weight of tissue used to produce the initial 50 µL of solution will allow a polyamine concentration to be determined.

Usual retention times in minutes for the polyamines are as follows: Putrescine, 11.0; cadaverine, 11.8; DAH, 13.8; spermidine, 16.5; aminopropylcadaverine, 16.8; spermine, 19.2; and *bis*(aminopropyl)cadaverine 19.2. As can be seen, this method will separate closely related polyamines, e.g., putrescine and cadaverine, as well as spermidine and aminopropylcadaverine, both pairs only differing in one methylene group. Unfortunately, this method will not separate spermine from its analog, *bis*(aminopropyl)cadaverine. Although the separation among other analogs is small, the peaks are sufficiently well-separated to quantify. However, it is advisable to spike an unknown sample with authentic standard to get initial peak assignments.

4. Notes

1. To determine free polyamines only, the sulfosalicylic acid method is the simplest. To the cell pellet, add 250 µL 8% 5-sulfosalicylic acid and 4 nmol DAH for every 10^6 mammalian cells and sonicate to resuspend the pellet. Hold on ice for 60 min. Centrifuge and remove supernatant. This can then be lyophilized and dansylated as described in **Subheading 3., steps 4–6**.
2. Alternatively, the cells can be treated with perchloric acid. Add 250 µL of water to resuspend the pellet, then add 250 µL 0.8*M* perchloric acid and mix. Add 20 nmol DAH. Hold on ice for 30 min. Centrifuge and remove supernatant. To the supernatant, carefully add 100 µL of 2*M* potassium carbonate. This precipitates the perchloric acid as potassium perchlorate, with the evolution of carbon dioxide. This precipitate is removed by centrifugation. Then 100 µL 5*M* HCl is carefully added to remove any residual carbonate. The solution is then lyophilized and dansylated as described **Subheading 3., steps 4–6**.
3. The derivatized polyamine solution can be purified using C_{18} 0.1 mL solid phase extraction columns. The columns are washed with two column volumes of methanol, followed by two column volumes of deionized water. The dansylated sample is then passed through the column and the unbound material washed out with two further volumes of water. When the column has drained, the dansylated polyamines are eluted with 500 µL of methanol. Once eluted, the derivatives are stable at 4°C for up to a month. This treatment slightly improves the chromatograms, but increases the storage time of the solution, possibly because of the removal of water and other impurities. The solid-phase extraction column used in the clean-up procedure can be used three times with no loss of efficiency, provided they are washed sufficiently with methanol before reuse.
4. Preparation of the cell extracts is relatively quick; several samples can be prepared in parallel within 90 min, the limiting step being the 30 min needed for protein precipitation at ice temperature. The lyophilization of the extracts prior to analysis usually takes several hours, but can be done overnight because the polyamine salts are stable for some time at room temperature. Lyophilization following acid hydrolysis can be carried out over approx 1 h. Derivatization of the lyophilized extracts in parallel takes approx 30 min. Each sample can be analyzed by HPLC in <30 min.
5. Hydrolysis can not be performed after the sample has been dansylated in this method, because the resultant charring causes peaks coeluting with the internal standard.
6. To remove excess underivatized dansyl chloride, amino acids other than proline can be used if necessary, but this does not improve the chromatogram and the removal is slower.
7. When derivatizing the samples at 70°C, it is advisable to use screw-capped vials because ordinary capped centrifuge tubes can open under the pressure of the boil-

ing acetone, causing a loss of the material. However, ordinary tubes are satisfactory if watched.

8. The sensitivity of the method is as follows: 10 pmol for putrescine and 5 pmol for spermidine and spermine.
9. The advantages of this method are that it is quick and convenient, the resolution between polyamines of very similar structure (e.g., spermidine and aminopropylcadaverine) is good, and it allows for easy inclusion of an internal standard to improve quantitation. On the negative side, once the sample is derivatized, it needs to be run quickly if clean-up is not used. Also, amino acids cannot be assayed in the same sample and a blank run is needed for quantitation and qualitation of small amounts of polyamines, making the assay slightly longer.

References

1. Kabra, P. M., Lee, H. K., Lubich, W. P., and Marton, L. J. (1986) Solid-phase extraction and determination of dansyl derivatives of unconjugated and conjugated polyamines by reversed-phase liquid chromatography: improved separation systems for polyamines in cerebrospinal fluid, urine and tissue. *J. Chromatogr.* **380,** 19–32.
2. Seiler, N. (1983) Liquid chromatographic methods for assaying polyamines using precolumn derivatization. *Methods Enzymol.* **94,** 10–25.
3. Westhall, F. and Hesser, H. (1974) Fifteen minute acid hydrolysis of peptides. *Anal. Biochem.* **61,** 610–613.
4. Hunter, K. J., Le Quesne, S. A., and Fairlamb, A. H. (1994) Identification and biosynthesis of N^1,N^9-*bis*(glutathionyl)aminopropylcadaverine (homotrypanothione) in *Trypanosoma cruzi*. *Eur. J. Biochem.* **226,** 1019–1027.

15

The Determination of Polyamines and Amino Acids by a Fluorescamine-HPLC Method

Karl J. Hunter and Alan H. Fairlamb

1. Introduction

The derivatization of polyamines and amino acids with fluorescamine is a useful tool for the detection and quantitation of these compounds. High-pressure liquid chromatography (HPLC) separation coupled to postcolumn derivatization allows for the detection of nanomole amounts of material and is simple to use. Because the compounds of interest are derivatized after separation by HPLC, their biological functionality is not destroyed prior to analysis, allowing the mixture to be separated for other uses. This method also has the advantage of being able to detect small peptides and conjugated polyamines (e.g., glutathione and trypanothione) and with the use of acid hydrolysis, allows for their composition to be analyzed.

2. Materials

1. Water-saturated ethyl acetate: Shake ethyl acetate and deionized water together, allow the mixture to settle, and use the top organic phase.
2. Propionic/hydrochloric acid: Propionic acid and 12*M* constant boiling hydrochloric acid (Pierce Ltd., Cambridge, UK) are carefully mixed in equal volumes. The mixture is stable at room temperature, but should be made fresh for each analysis.
3. Lithium hydroxide: 2*M* aqueous solution. On storage the solution will go slightly cloudy, but can still be used. Keep in a tightly stoppered bottle.
4. The two HPLC solvents are solvent A: 0.25% (w/v) lithium camphorsulfonate, pH 2.65; solvent B: 0.25% (w/v) lithium camphorsulfonate, pH 2.65, in 25% (v/v) 1-propanol. (In our experience, 1-propanol from Romil Chemicals [Loughborough, UK] gives the best results.)

   A stock solution of 1% (1S)-(+)-10-camphorsulfonic acid (CSA) is prepared by dissolving 20 g in 1.8 L and adjusting the pH to 2.0 with 2*M* lithium hydroxide

From: *Methods in Molecular Biology, Vol. 79: Polyamine Protocols*
Edited by: D. Morgan Humana Press Inc., Totowa, NJ

solution. After making up to 2 L with water, it is important to recheck that the pH is 2.0. The pH meter should be equipped with a glass electrode and calibrated with an appropriate low pH standard solution. For HPLC, dilute this stock solution 1 in 4 for solvent A, mix, and pass through a 0.45-μm filter. For solvent B, mix 1% CSA:1-propanol:water (1:1:2). Filter as before.

5. Fluorescamine: This is very sensitive to both amines and water, so acetone must be redistilled over ninhydrin before use as a solvent. Distillation should be carried out in an efficient fumehood. To avoid the distilled acetone reabsorbing water, the outlet of the condenser is joined to the collecting bottle via a Quickfit joint, with an air outlet passing through a silica gel trap. 1.5 L of GPR grade acetone, approx 1 g of Ninhydrin, and a few boiling chips are put into the flask and the solution is heated on an electric mantle until the vapor temperature reaches 56°C, then the heat is turned down. The first 100 mL of solution is discarded, together with the final 100 mL. The acetone is stored in air-tight bottles until mixed with fluorescamine (200 mg/L) for use in postcolumn derivatization. The container for postcolumn derivatization also has an air inlet containing a silica gel trap.
6. 0.2*M* lithium borate solution: Crystalline boric acid is dissolved in 1.8 L of deionized water and 2*M* lithium hydroxide solution is added (approx 100 mL) to a pH of 9.5. The solution is made up to 2 L and the pH rechecked. For use, the solution is diluted 1:1 in deionized water and filtered through a 0.45- or 0.22-μm filter.
7. DAH internal standard: 1,7-Diaminoheptane (DAH) was converted to the nonhygroscopic dihydrochloride by dissolving in excess 1*M* hydrochloric acid, followed by lyophilization. DAH is dissolved in solvent A at a concentration of 400 μ*M* and stored frozen prior to use. This solution can be stored at –20°C for several months.
8. Standard polyamine solution: A stock solution of putrescine dihydrochloride (100 μ*M*), spermidine trihydrochloride (800 μ*M*), and spermine tetrahydrochloride (800 μ*M*), are dissolved in solvent A. Any other required polyamines are made up similarly. This solution can be stored at –20°C for several months.
9. Standard amino acid solution: A suitable dilution of the Sigma amino acid standard solution (AAS-18) in solvent A. This solution can be stored at 4°C for several months.
10. An HPLC system equipped with a 5-μm ODS reversed-phase column (a 4.6 × 250 mm ion-paired column from Beckman Instruments Ltd. [High Wycombe, UK] works well), preferably with a guard column (e.g., Newguard RP-18 guard column from Anachem Ltd. [Luton, UK]) coupled to an in-line fluorescence detector (*see* **Note 1**).
11. All reagents should be of the highest purity available because the method is very sensitive to minor contaminants. All aqueous solutions are made up in freshly prepared deionized water (18 MΩ/cm).

3. Methods

1. Spin down the cells to be analyzed in a 1.5-mL microcentrifuge tube, wash 2–3 times in an appropriate buffer, e.g., phosphate-buffered saline for mammalian cells or salt solution, 0.9% (w/v) sodium chloride for bacteria.

2. Add 250 µL of water and resuspend the pellet, then add 250 µL 10% trichloroacetic acid dissolved in 0.01*M* HCl and mix. (The hydrochloric acid prevents polyamines from binding to glass as the pH increases). Add 10 µL of the DAH internal standard (400 µ*M* 1,7-diaminoheptane). Leave on ice for 30 min to allow the protein to precipitate. Centrifuge and remove the supernatant. Extract the supernatant five times with an equal volume of water-saturated ethyl acetate to remove excess trichloroacetic acid. If the free polyamine content only is required, the solution is then lyophilized. The residue can be redissolved in solvent A with sonication if necessary.
3. If the total polyamine content (free and conjugated) is required, after ethyl acetate extraction the solution is transferred to a reactivial (Pierce Ltd.) and lyophilized. The residue is then hydrolyzed ***(1)***. Add 50 µL propionic/hydrochloric acid to the reactivial and heat at 160°C in a heating block for 15 min. Allow to cool and lyophilize. Redissolve in solvent A with sonication. Filter through a 0.45-µm syringe filter.
4. Inject 20 µL onto the HPLC. The running conditions are as follows:
   a. Temperature: ambient
   b. Elution gradient on injection: 0% B to 20% B over 60 min.
      at 60 min: 20% B to 75% B over 40 min.
      at 100 min: 75% B to 0 % B over 1 min (stop data).
      at 140 min: End method (next sample injected).
   c. Detection: Fluorescence following postcolumn derivatization (*see* **Note 1**), 390 nm (excitation), 480 nm (emission).
   d. Calibration: A standard is assayed. This consists of a solution of 0.5 nmol/20 µL injected putrescine and 4 nmol/20 µL injected spermidine and spermine in solvent A. (Other diamines and polyamines can be added at the same concentrations). For amino acids, the standard solution is diluted 100-fold into the polyamine standard. Internal standard is added to the solutions at a concentration such that 1 nmol DAH is injected per sample.

Because the amount of polyamine or amino acid in the test solution and the amount lost via experimental manipulation (from the difference in the DAH peak) are known, the true concentrations in the test solution can be determined. Knowing the number of cells or weight of tissue used to produce the initial solution will allow a concentration of compound per cell can be determined.

It is also recommended that a reagent blank (solvent A alone) be run when the system is set up and at regular intervals. Because the method is very sensitive to amine-containing contaminants, a blank will allow for the detection of spurious peaks that can then be subtracted from subsequent chromatograms.

4. Notes

1. A general scheme of an HPLC equipped with an on-line postcolumn detection system is illustrated in **Fig. 1.** The sample eluted from the HPLC column first passes through a UV-detector that can be used to quantitate metabolites associ-

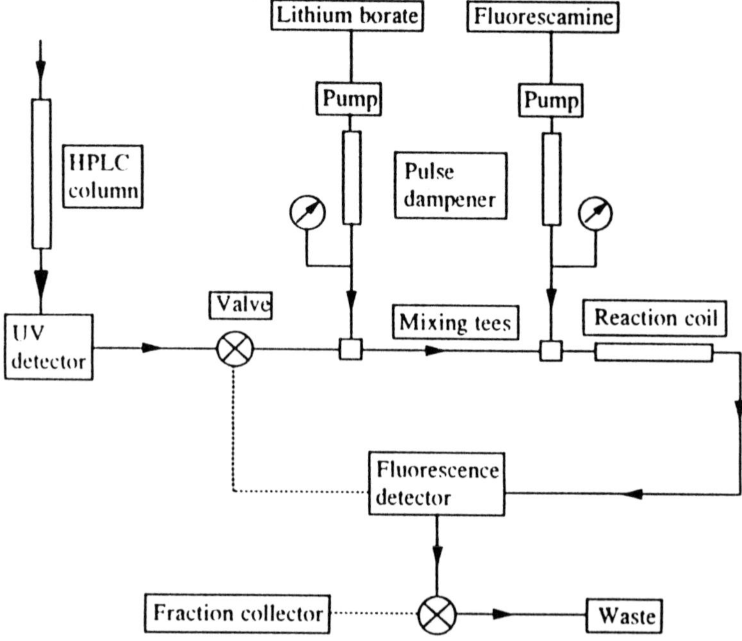

Fig. 1. Schematic diagram of the HPLC postcolumn detection system.

ated with polyamine metabolism, such as adenosylmethionine and decarboxylated adenosylmethionine, or other peptides, drugs, or metabolites. The effluent is then mixed with the lithium borate buffer to raise the pH of the HPLC solvent mixture to the optimum (pH 8.0–9.0) for reaction with fluorescamine. A short delay coil allows the reaction of the fluorescamine with primary amine groups to go to completion before passing through a fluorescence detector (Gilson Model 121 Fluoromonitor fitted with 310–410 nm excitation and 475–650 nm emission filters). A valve allows the system to be bypassed when the postcolumn reactor is not in use, minimizing peak broadening for other analyses that can be performed with the same solvent system. A second valve allows the effluent to be collected for further analysis, such as scintillation counting. The postcolumn pumps should be single piston pumps capable of accurate low flow rates and resistant to acetone and alkaline borate buffer solutions. Milton Roy (Riviera Beach, FL) duplex minpumps capable of delivering 0.22–4.4 mL/min are ideal for this purpose. Our current system uses 0.508 mm id stainless steel tubing except for the pulse-dampening coils, which are approx 0.30 m of coiled 0.635 mm id Teflon tubing (the lithium borate causes unacceptable corrosion of commercially available stainless steel pulse-dampeners). The reaction coil comprises 0.30 m of tubing, which introduces a delay of approx 20 s between mixing of the reagents and detection of fluorescence. The mixing tees are Visco Jet micromixers from the Lee Company (Westbrook, CT). System pressure is monitored by conventional 0–20 bar pressure gages. The reagent reservoir tubing (1.575 mm id Teflon) should be fitted with taps and bubble traps, since fluorescamine is virtually insoluble in water and can form precipitates when the system is not running. A pressure relief valve set at about 10 bar is useful in avoiding serious damage to the plumbing connections should a blockage occur. We have not found a back pressure regulator necessary with our current system.

To optimize conditions, set the HPLC flow rate at 1 mL/min with 50% solvent B and then adjust the flow of borate buffer (usually 0.4 mL/min) so the pH of the effluent is between 8.0 and 9.0. The fluorescamine pump is then switched on at about 0.4 mL/min, a standard amount (1 nmol) of nor-leucine is injected into the column, and the fluorescence response recorded. Minor adjustments to the flow rate through each pump are then made to obtain the optimal response.

Once constructed, the detection system is simple and easy to use. It is as sensitive and reproducible as the other commonly used reagent (*o*-phthaladehyde) for the detection of amino acids and polyamines, but has the added advantage of being more sensitive in the detection of polypeptides (e.g., trypanothione). Postcolumn methods are also useful for the purification of amine-containing molecules with biological activity, since they can be coupled to a stream-splitting device (or switched off during a preparative run). In contrast, precolumn derivatization inevitably destroys biological activity. The current solvent system is also flexible and has been used in our laboratory for the separation and quantitation of arsenical drugs ***(3,4)*** and their biological adducts ***(5–7)***, difluoromethylarginine and other guanidine-containing compounds ***(8)***, peptide metabolites ***(9,10)***, purine metabolites ***(11)***, and thiols ***(10–12)***.

2. Peak assignments are made by the injection of each individual amino acid or polyamine. Also, spiking an unknown sample with authentic standards is recommended. This method will separate closely related polyamines, e.g., putrescine and cadaverine, but not spermidine and aminopropylcadaverine, or spermine from its analog, *bis*(aminopropyl)cadaverine. The amino acid standard contains a mixture comparable with acid-hydrolyzed protein. Therefore, it does not contain asparagine, glutamine, or tryptophan; however, these acids contain a primary amino group and can be detected by this system if added in. (Acid hydrolysis destroys tryptophan in samples ***[1]***). Because of its ease of oxidation, cysteine is also not included in the amino acid mixture, being substituted by cystine, but can be detected if added. Proline is contained in the amino acid mixture, but cannot be detected by this system because it does not contain a primary amino group, being an imino acid. (A method is available for the conversion of proline to an amino acid, which can be detected in this system ***[2]***). The separation of the amino acids is generally good, although alanine, aspartate, glutamate, and glycine tend to elute close to the solvent front and the peaks overlap to some extent.
3. Preparation of the cell extracts is relatively quick; several samples can be prepared in parallel within 90 min, the limiting step being the 30 min needed for protein precipitation at ice temperature. The lyophilization of the extracts prior to

analysis usually takes several hours, but can be done overnight because the polyamine and amino acid salts are stable for some time at room temperature. Lyophilization following acid hydrolysis can be carried out over approx 1 h. The HPLC analysis takes 100 min/sample.

4. The sensitivity of the method is approx 50 pmol/injection for putrescine, 100 pmol/injection for the amino acids and agmatine, and 400 pmol/injection for spermidine and spermine.

References

1. Westhall, F. and Hesser, H. (1974) Fifteen minute acid hydrolysis of peptides. *Anal. Biochem.* **61,** 610–613.
2. Weigele, M., Debernado, S., and Leimgruber, W. (1973) Fluorometric assay of secondary amino acids. *Biochem. Biophys. Res. Commun.* **50,** 352–356.
3. Berger, B. J. and Fairlamb, A. H. (1988) High performance liquid chromatographic method for the separation and quantitative estimation of anti-parasitic melaminophenyl arsenical compounds. *Trans. Roy. Soc. Trop. Med. Hyg.* **88,** 357–359.
4. Berger, B. J. and Fairlamb, A. H. (1994) Properties of melarsamine hydrochloride (Cymelarsan) in aqueous solution. *Antimicrob. Agents Chemother.* **38,** 1298–1302.
5. Fairlamb, A. H., Henderson, G. B., and Cerami A. (1989) Trypanothione is the primary target for arsenical drugs against African trypanosomes. *Proc. Natl. Acad. Sci. USA* **86,** 2607–2611.
6. Fairlamb, A. H., Smith, K., and Hunter, K. J. (1992) The interaction of arsenical drugs with dihydrolipoamide and dihydrolipoamide dehydrogenase from arsenical resistant and sensitive strains of *Trypanosoma brucei brucei. Molec. Biochem. Parasitol.* **53,** 223–232.
7. Cunningham, M. L. and Fairlamb, A. H. (1995) Trypanothione reductase from *Leishmania donovani*—purification, characterisation and inhibition by trivalent antimonials. *Eur. J. Biochem.* **230,** 460–468.
8. Hunter, K. J. and Fairlamb, A. H. (1990) Separation and quantitation of the polyamine biosynthesis inhibitor DL-α-difluoromethylarginine and other guanidine-containing compounds by high-performance liquid chromatography. *Anal. Biochem.* **190,** 281–285.
9. Fairlamb, A. H., Blackburn, P., Ulrich, P., Chait, B. T., and Cerami, A. (1985) Trypanothione: a novel *bis*(glutathionyl)spermidine cofactor for glutathione reductase in trypanosomatids. *Science* **227,** 1485–1487.
10. Hunter, K. J., Le Quesne, S. A., and Fairlamb, A. H. (1994) Identification and biosynthesis of N^1,N^9-*bis*(glutathionyl)aminopropylcadaverine (homotrypanothione) in *Trypanosoma cruzi. Eur. J. Biochem.* **226,** 1019–1027.
11. Fairlamb, A. H, Henderson, G. B., Bacchi, C. J., and Cerami, A. (1987) *In vivo* effects of difluoromethylornithine on trypanothione and polyamine levels in bloodstream forms of *Trypanosoma brucei. Mol. Biochem. Parasitol.* **24,** 185–191.
12. Shim, H. and Fairlamb, A. H. (1988) Levels of polyamines, glutathione and glutathione-spermidine conjugates during growth of the insect trypanosomatid *Crithidia fasciculata. J. Gen. Microbiol.* **134,** 807–817.

16

Thin-Layer Chromatographic Method for Assaying Polyamines

Rentala Madhubala

1. Introduction

The naturally occurring polyamines putrescine, spermidine, and spermine are found in all tissues and microorganisms *(1–5)*. Polyamines are cations that have been implicated in various growth processes and in cellular differentiation *(6)*. This chapter describes a simple and rapid method for the assay of polyamines. The method described here is based on that of Seiler *(7)* and enables one to determine 10–100 pmol of polyamines.

The natural polyamines cannot be detected by optical or electrochemical methods. Prechromatographic derivatization of polyamines with reagents that lead to the formation of derivatives having high molar extinction coefficients or fluorescence has been successfully used for their detection by suitable electrochemical detectors. The basic principle used here for the isolation of polyamines is by prechromatographic derivatization using an acid chloride, 5-dimethylaminonapthalene-1-sulfonyl chloride (dansyl chloride). The advantages of prechromatographic derivatization are twofold: First, it ensures the solubility of the derivatives in organic solvents; enabling their extraction from aqueous phases. Second, these stable derivatives can be isolated by chromatography. This method also facilitates the determination of radioactivity with great precision if labeled precursors are used for the metabolic studies. The derivatized polyamines or dansyl derivatives can then be used for separation by thin layer chromatography and subsequent evaluation by *in situ* scanning of fluorescence or by extraction followed by conventional measurement of fluorescence.

The thin layer chromatography method described in this chapter may not be as specific and precise as other methods used for polyamine assay, but it still helps in the detection of polyamines by a simple and rapid means without

From: *Methods in Molecular Biology, Vol. 79: Polyamine Protocols*
Edited by: D. Morgan Humana Press Inc., Totowa, NJ

involving the usage of any sophisticated equipment. It also enables one to perform several separations in parallel.

2. Materials

1. Tissue or cultured cells.
2. Dulbecco's phosphate-buffered saline (PBS) "A".
3. Perchloric acid (2%): Dissolve 2 mL of perchloric acid in 100 mL of deionized water. Store at room temperature.
4. Supersaturated sodium carbonate. Store at 4°C.
5. Dansyl chloride: Dissolve 5 mg of dansyl chloride (Sigma [St. Louis, MO] D-2625) in 1 mL of acetone. This solution should be prepared fresh just before use and should be used in the dark because it is light sensitive (*see* **Note 1**).
6. L-Proline solution: Dissolve 150 mg of proline in 1 mL of deionized water. This should be prepared fresh just before use.
7. Toluene.
8. Cyclohexane:ethyl acetate (2:3 v/v).
9. Standard polyamine solutions (1*M* each). Dissolve 161.1 mg of putrescine (Sigma-P-7505), 254.6 mg of spermidine (Sigma-S-2501), and 348.2 mg of spermine (Sigma-S-2876), respectively, in 1 mL of deionized water. Dilute the stocks to 1 m*M* each in deionized water. Store the stocks at 4°C.
10. 0.20-mm thick silica gel on precoated plates of size 20 × 20 cm. Store the plates in a dark place and keep protected from chemical vapors.
11. Tissue homogenizer.
12. Thin-layer chromatography scanner (TLC Scanner II/CAMAG with CATS3/TLC II software program).
13. Hamilton syringe (100 mL) specially manufactured for Camag equipment (Switzerland).
14. Linomat IV: It is an automatic device to load the samples on the TLC plates.
15. UV cabinet.
16. Glass tank with lid for developing TLC plates.
17. Spectrofluoro-Photometer (Shimadzu).

3. Methods

3.1. Extraction of Polyamines

3.1.1. Tissue Extracts

Tissue samples weighing approx 0.1 g of the tissue are washed with PBS and homogenized using 10 vol of 2% (v/v) perchloric acid. Homogenization should be done at 4°C.

3.1.2. Cell Extracts

1. Harvest 10^6–10^7 cells. Wash with PBS. Sonicate the cells in 250 μL of 2% (v/v) perchloric acid. Sonication should be performed at 4°C.

2. Tissue homogenate and cell homogenates obtained are then stored at 4°C for 24 h. After 24 h the samples are centrifuged at approx 11,500*g* for 30 min at 4°C.
3. The supernatant is carefully collected and stored at 4°C. This is then used for estimating polyamines.

3.2. Dansylation

1. Routinely 200 μL of the supernatant is used for dansylation (*see* **Note 2**).
2. External standards with known amount of polyamines are also used for dansylation.
3. Samples with internal standards are prepared by adding to the tissue or cell supernatants known amount of polyamines, corresponding approximately to the amount of polyamines expected in the samples.
4. Blank containing the appropriate amount (200 μL) of perchloric acid is also prepared.
5. In order to dansylate the samples, add twice the sample volume (400 μL) of a solution of dansyl chloride in acetone (*see* **Note 3**).
6. Add 200 μL of supersaturated sodium carbonate solution to this. Vortex and store in the dark at room temperature for 16 h (*see* **Note 4**).
7. Excess of dansyl chloride is removed at this stage by adding 100 μL of proline solution. Store the samples in the dark for another 30 min.
8. Extract the dansyl amides with 500 μL of toluene by vortexing the tubes at room temperature. Separate the aqueous phase from the toluene phase by centrifugation. The toluene phase that contains the dansylamides is now ready for separation by TLC.

3.3. Thin Layer Chromatography

3.3.1. Separation of Dansyl Amides

1. Apply the samples in small spots to the thin layer plate by using Linomat IV (CAMAG) with a Hamilton syringe specially designed for Camag equipments. Linomat IV is microprocessor-controlled and programmable and enables one to assign samples to available tracks and choose the desired volume. The track location is automatic along with several sample deliveries from the same syringe. The samples are applied in the size of 6-mm spots with a distance of 18 mm from one edge. The distance between the neighboring spots is 6 mm. Sixteen samples can be loaded on a plate of size 20 × 20 cm (*see* **Notes 5** and **6**).
2. The solvent cyclohexane:ethylacetate (2:3) is placed in the chromatography tank for saturation.
3. The plates are now exposed to ascending chromatography in the solvent saturated glass tank. The plates are immersed in the solvent mixture for 1.5–2 h or until the solvent front reaches just 2.5–5 cm below the end mark.
4. Remove the plates from the tank and allow to dry in air in the dark (*see* **Note 7**).

3.3.2. Quantitative Evaluation

1. This can be achieved either by *in situ* scanning of fluorescence measurement or after extraction by conventional fluorescence measurement (*see* **Notes 8** and **9**).
2. In order to evaluate polyamines by *in situ* scanning, CATS software has been developed to control data recording and evaluation with CAMAG Scanner II

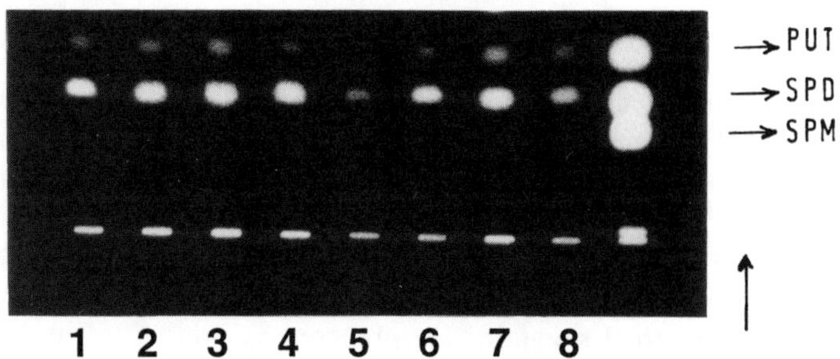

Fig. 1. Separation pattern of dansylated putrescine, spermidine, and spermine on a precoated silica get plate using ethylacetate:cyclohexane (2:3) as the solvent system. The lane at the extreme right shows a separation pattern of standard polyamines. The arrows indicate: putrescine (PUT), spermidine (SPD), and spermine (SPM) bands. Lanes 1–4 show polyamine separation in the strain UR-6. Lanes 5–8 show polyamine separation in the strain AG-83. UR-6 and AG-83 are strains of *Leishmania donovani*.

(Switzerland). TLC plates are examined in the UV cabinet **(Fig. 1)** and then scanned with light (UV or VIS) of selected wavelength. The data from unknown (samples) and standard tracks are stored. The integration program is then used to calculate the peak heights and peak areas **(Fig. 2)**. This data is also stored. In the calibration program the peak data is converted into quantitative units for each identified component and an analytical record containing a statistical evaluation is printed out. Absorption spectra recording permits automatic testing of separated substrates for identity and purity.

3. If conventional fluorimetry is the method of choice for quantitative evaluation of dansyl derivatives then the derivatives are first eluted from the spots. After chromatography the TLC plates are visualized with the aid of 360 nm UV source. The zones containing the dansyl derivatives are scraped off with a scalpel blade and adsorbent is collected in a 10 mL centrifuge tube. The dansyl polyamines are extracted with 10 mL quantities of toluene and assayed in a conventional spectrofluorometer at excitation 365 nm and emission 505 nm. The fluorescence corresponding to each spot is measured and compared with that of a known standard. The concentrations of sample polyamines are determined with standards and normalized per milligram of sample protein. Radioactive polyamines can also be analyzed as dansyl derivatives in metabolic studies. In this case the TLC plates are apposed to films (Kodak) for 2–4 d. Spots are identified by autoradiography, scraped into vials containing 10 mL of Liquid Scintillation fluid ***(8)***, and counted in a Scintillation counter (Packard).

4. Notes

1. Dansyl chloride is purchased from Sigma (catalog no.D-2625). It is relatively pure and does not require further purification. However, impure samples of dansyl chloride can be purified according to Seiler ***(7)***: 10 g of product is dissolved in

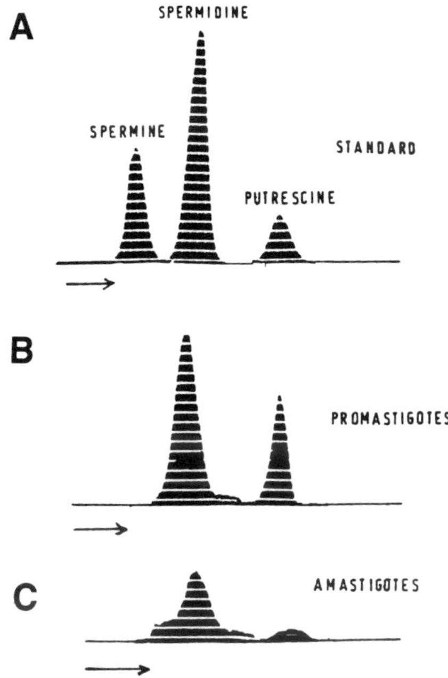

Fig. 2. Representative densitometer fluorescent tracings of TLC plates using TLC scanner II/CAMAG with CATS 3/TLC II software program. **(A)** Polyamine standards. **(B)** Polyamines in promastigote culture of *Leishmania donovani* (1×10^7 cells). **(C)** Polyamines in amastigote culture of *Leishmania donovani* (1×10^7 cells). The arrows indicate the direction of the densitometer scan.

20 mL distilled toluene. The solution is then passed through a silica gel 60 (70–230 mesh) column (12–100 mm). Dansyl chloride is then eluted from the column with toluene. The impurities get adsorbed to the silica gel. The elute is dried in vacuum to get deep orange crystals (melting point 69°C).

2. Prechromatographic derivatization of polyamines with acid chlorides, specifically dansyl chlorides, is used more commonly than fluorescamine because the latter is a more specific reagent that reacts only with primary amino groups, in contrast to dansyl (acid) chloride, which forms derivatives not only with primary and secondary amino groups, but also with imidazole nitrogen, phenolic hydroxyls, and even some alcohols *(6)*.
3. Prechromatographic derivatization reagents are added in excess of quantity in order to obtain the reaction with all amino groups of the polyamines.
4. Dansyl derivatives are light sensitive, hence the reactions and running of TLC plates should be done in the dark.
5. The critical point to be taken care of for running TLC is sample application. In our experience use of a fully automated device, such as Linomat IV from CAMAG (Switzerland), yields reproducible results.

6. The commercially supplied TLC plates (Anchrom, India) did not require activation for 1 h at 110°C before use.
7. Since dansyl derivatives are light-sensitive and can be irreversibly destroyed on active surfaces by light, it is essential to keep the dry plates in the dark until they are ready for quantitative determination.
8. It was suggested earlier that spraying the plates with 20 mL of a solution of trietanol amine in 2-propanol (1:4) stabilized the dansyl derivatives and increased fluorescence. The commercially available plates obtained now do not require this step and we have found that fluorescence intensity is quite stable for 3–4 d if they are stored in the dark.
9. The results obtained from *in situ* and elution methods enable one to determine 10–100 pmol of polyamines.

References

1. Heby, O. (1981) Role of polyamines in the control of cell proliferation and differentiation. *Differentiation* **19,** 1–20.
2. Heby, O., Marton, L. J., Wilson, C. B., and Martinez, H. M. (1975) Polyamines: a high correlation with cell replication. *FEBS Lett.* **50,** 1–4.
3. Tabor, C. W. and Tabor, H. (1984) Polyamines. *Ann. Rev. Biochem.* **53,** 749–790.
4. Pegg, A. E. and McCann, P. P. (1982) Polyamine metabolism and function. *Am. J. Physiol.* **243,** C212–C221.
5. Pegg, A. E. (1986) Recent advances in the biochemistry of polyamines in eukaryotes. *Biochem. J.* **234,** 249–262.
6. Igarashi, K., Hara, K., Watanbe, Y., and Takeda, Y. (1975) Polyamine and magnesium contents and polypeptide synthesis as a function of cell growth. *Biochem. Biophys. Res. Commun.* **64,** 897–904.
7. Seiler, N. (1983) Thin layer chromatography and thin layer electrophoresis of polyamines and their derivatives. *Methods Enzymol.* **94,** 3–25.
8. Bray, G. A. (1960) A simple and efficient liquid for counting aqueous solutions in a liquid scintillation counter. *Anal. Biochem.* **1,** 279–285.

V

Measurement of Polyamine Transport

17

Measurement of Polyamine Transport

Adherent Cells

David M. L. Morgan

1. Introduction

Most cells can take up polyamines, and some polyamine analogs, by a process that is independent of amino acid transport systems *(1,2)*, and that, in certain circumstances, can substitute for synthesis *de novo*. In some cells polyamines are taken up by both saturable and nonsaturable systems. Uptake by saturable systems is energy-requiring, temperature-dependent, carrier-mediated, and operates against a substantial concentration gradient *(3,4)*. In many cells putrescine and, often, spermidine transport is sodium-sensitive (the term sodium-sensitive is preferable to the more frequently used sodium-dependent, because removal of extracellular sodium reduces, but often does not abolish, uptake). In contrast, spermine transport in most cells is not affected by changes in extracellular sodium concentrations. The number of carriers in the polyamine transport system varies with cell type. In human umbilical vein endothelial cells there are apparently two carriers, one shared by spermine and spermidine, and one capable of transporting all three polyamines *(5)*. In contrast, porcine aortic endothelial cells appear to possess three carriers, one for each polyamine *(6)*. Multiple carriers are not uncommon, although many cell types appear to possess a single polyamine carrier able to transport putrescine, spermidine, and spermine, although with differing affinities *(1)*.

The mechanisms of polyamine transport and the means by which polyamine uptake is induced are not clear. Uptake is generally low in quiescent cells, or in cells that have been induced to differentiate, in contrast to cells in rapid growth in which uptake is enhanced. Uptake is also increased in response to proliferative stimuli, such as serum, growth factors, and hormones, and in cells in which

From: *Methods in Molecular Biology, Vol. 79: Polyamine Protocols*
Edited by: D. Morgan Humana Press Inc., Totowa, NJ

polyamine stores have been depleted, for example, by the use of inhibitors of polyamine biosynthesis, such as difluoromethylornithine ***(6–9)***, an enzyme-activated inhibitor of ornithine decarboxylase ***(10)***. It seems likely ***(11,12)*** that polyamine transport is regulated by both a rapidly degraded inhibitory protein, which responds to a rise in intracellular polyamine levels, and a longer lived protein, or proteins, which may be the polyamine carrier(s).

The procedure is described here for use with the human umbilical vein endothelial cell line ECV304 ***(13)***, obtainable from cell culture collections. It has also been used with freshly isolated human umbilical vein endothelial cells ***(5)***, human umbilical artery smooth muscle cells, and J774 cells (a mouse macrophage cell line), but can be applied equally well to any adherent cells grown in monolayer cultures. The polyamine concentrations are within the physiological range and the initial cell concentration is such that the cells are in continuous growth when used. The protocol describes the simultaneous use of six different polyamine concentrations, each replicated 10 times, on a single 96-well microplate. It can easily be adapted to the use of 10 different concentrations, with six replicates; <2% of the applied label is taken up (*see* **Note 1**).

2. Materials

1. Flat-bottomed 96-well tissue culture grade microwell plates with lids (e.g., 1-67008A, Life Technologies Ltd., [Paisley, Scotland]; 655180, Greiner [Frickenhausen, Germany]; 76-003-05, ICN Pharmaceuticals Inc. [Thame, UK, and Costa Mesa, CA]).
2. Repetitive pipet, e.g., Multipet (Eppendorf 4780, Merck Ltd. [Lutterworth, UK]).
3. Combitip, 5 mL, for Multipet (e.g., 307/8534/03, Merck Ltd; NP7331, Alpha Laboratories [Eastleigh, UK])
4. Multichannel pipets (2), e.g., Proline 8-channel (Alpha Laboratories).
5. 8-Channel repetitive dispenser (e.g., BCL 8800, BCL Corporation [Lewes, UK]).
6. Syringes for 8-channel repetitive dispenser (88-SBR, BCL Corporation).
7. 8-Well reagent reservoir, polypropylene (78-519-02, Flow Laboratories Ltd.), or vial carrier, 12-place (76-720-00, Flow Laboratories Ltd.) and reagent vials, 1-mL (76-717-07, Flow Laboratories Ltd.) (*see* **Note 2**).
8. Reagent trough (77-824-01, Flow Laboratories Ltd.).
9. Ice tray, 25 × 30 × 5 cm.
10. Ice bucket, capable of cooling a 500 mL reagent bottle.
11. Eppendorf tubes.
12. Polypropylene mini scintillation vials (LIP Ltd. [Shipley, UK]).
13. Trays for scintillation vials (KH44/16, Denley Instruments Ltd. [Billingshurst, UK].
14. Optiphase "Safe" scintillation fluid (Wallac UK Ltd. [Milton Keynes, UK]).
15. Hank's balanced salt solution (HBSS).
16. 1 *M* HEPES solution: Dissolve 47.6 g of HEPES (Ultrol grade, Calbiochem [San Diego, CA]) in 200 mL tissue-culture-grade deionized water, sterilize by filtration through a 0.2-μm filter, and store as 5 mL aliquots in sterile plastic bijous at –20°C. Shelf life ~6 mo.

17. HBSS modified by additional buffering with HEPES (MHBSS; *see* **Note 3**); under sterile conditions add 5 mL of 1*M* HEPES to 500 mL of HBSS and adjust the pH to 7.4 using 1*M* NaOH. Store at 4°C; and use within 4 wk.
18. Phosphate-buffered saline, Dulbecco's "A" (PBS).
19. 10 m*M* Spermine solution; dissolve 17.4 mg spermine tetrahydrochloride in 5 mL tissue-culture-grade deionized water, sterilize by filtration through a 0.2-μm filter, and store at 4°C. Shelf life ~6 mo.
20. 10 m*M* Spermidine solution; dissolve 12.7 mg spermidine trihydrochloride in 5 mL tissue-culture-grade deionized water, sterilize by filtration through a 0.2-μm filter, and store at 4°C. Shelf life ~6 mo.
21. 10 m*M* Putrescine solution; dissolve 8 mg putrescine dihydrochloride in 5 mL tissue-culture-grade deionized water, sterilize by filtration through a 0.2-μm filter, and store at 4°C. Shelf life ~6 mo.
22. 100 μ*M* solutions of [^{3}H]putrescine ([1,4(n)-^{3}H]tetramethylenediamine dihydrochloride, 22.9 Ci/mmol, 1 mCi/mL; Amersham International plc [Amersham, UK]), or [^{3}H]spermidine (*N*-(3-[^{3}H]aminopropyl)-1,4-[^{3}H]diamine trihydrochloride, 17.6 Ci/mmol, 1 mCi/mL; Dupont-NEN [Stevenage, UK]) are prepared by adding 50 μL of radiolabeled material to 50 μL of 10 m*M* unlabeled polyamine and diluting to 5 mL with MHBSS (final specific activity 50–100 mCi/mmol). Store at 4°C and use within 4 wk.
23. 100 μ*M* [^{14}C]spermine is prepared by simple dilution of [^{14}C]spermine (*N,N'*-bis-(3-aminopropyl)-[1,4-^{14}C]tetramethylene-1,4-diamine tetrahydrochloride, 118 mCi/mmol, 50 μCi/mL; Amersham International plc); because of the low specific activity, no addition of unlabeled material is necessary. Store at 4°C and use within 4 wk.
24. Working solutions of radiolabeled polyamine (0.25, 0.5, 1.0, 2.5, 5.0, and 10.0 μ*M*) are prepared (1 mL of each) in 1.5 mL Eppendorf tubes by dilution of the 100 μ*M* stock solutions with MHBSS, as follows.

100 μ*M* polyamine, μL	MHBSS, μL	Final concentration of polyamine, μ*M*
2.5	997.5	0.25
5.0	995	0.5
10.0	990	1.0
25.0	975	2.5
50.0	950	5.0
100.0	900	10.0

3. Methods

3.1. Subculture of Endothelial Cells

Cultures of ECV304 cells maintained in 75 cm^2 tissue culture flasks in medium 199 supplemented with 10% fetal calf serum, 100 U/mL penicillin and streptomycin, 21 m*M* bicarbonate, and 2 m*M* glutamine, are subcultured weekly.

1. Warm all solutions to 37°C.
2. Remove the medium from a confluent monolayer of cells and rinse the cells with serum-free medium (serum contains a trypsin inhibitor).
3. Add 2.5 mL of trypsin (0.1%)-EDTA (0.02%) solution in PBS to a 75 cm^2 flask and tilt the flask to thoroughly wet the cell monolayer. Return the flask to the incubator.
4. After 5 min remove the flask from the incubator and examine under the microscope. The cells should have rounded up and begun to detach from the surface. Smartly tap the side of the flask 2–3 times to aid cell detachment. Prolonged exposure to trypsin can damage endothelial cells and reduce viability.
5. Add an equal volume of serum-containing medium and draw the cell suspension up and down 3–4 times through a narrow bore pipet to break up any cell clumps. Divide the cell suspension among fresh 75 cm^2 flasks (split ratio 1:3–1:5, as required) and add sufficient serum-containing medium to bring the volume to 10 mL.
6. Return the cells to the incubator and feed every 2–3 d with fresh medium.

The subcultured cells should again become confluent in within 5–7 d.

3.2. Plating of Cells for Transport Experiments

1. Detach the cells from the required number of confluent cultures as described in **Subheading 3.1.**
2. Pool the cells and dilute in fresh growth medium to 5×10^4 cells/mL.
3. Using a Multipet equipped with a 5 mL Combitip, draw up some of the cell suspension. Set the Multipet to deliver 200 μL and eject the first aliquot back into the parent cell suspension (after filling, the first ejection may not deliver the full 200 μL).
4. Dispense 200 μL, approx 10^4 cells, into the inner 60 wells of a 96-well flat-bottomed tissue culture plate (cell growth is less reproducible in the outer wells).
5. Place 200 μL of PBS in the outer 36 wells.
6. Return the plates to the incubator and allow the cells to adhere for 24 h before use.

Techniques for the isolation, culture, and plating of human umbilical vein endothelial cells (HUVEC) have been described elsewhere ***(14)***.

3.3. Measurement of Polyamine Uptake

Steps 4–10 are carried out at 37°C; in **steps 10–18** timing is critical.

1. Warm MHBSS and the required polyamine solutions to 37°C.
2. Fill ice bucket and ice tray with melting ice (add water if necessary to ensure good thermal contact with items placed therein).
3. Place a bottle with sufficient PBS in the ice bucket to cool.
4. Pour some MHBSS into a reagent trough.
5. Dispense the polyamine solutions into individual reagent vials and place the vials in a vial carrier; alternatively, the polyamine solutions may be placed in individual channels of an 8-well reagent reservoir.

6. Draw up the polyamine solutions into six channels of a repetitive dispenser (*see* **Note 4**).
7. Set an alarm timer to 15 min (*see* Note 5), but do not start it yet.
8. Remove the medium from each well using a multichannel pipet (set the pipet to ~225 µL to ensure removal of all the medium in one operation); *see* **Notes 6** and **7**.
9. Add ~180 µL MHBSS to each well to rinse the cells, using a second multichannel pipet.
10. Remove the MHBSS using the multichannel pipet that has been set to 225 µL.
11. Using the multichannel repetitive dispenser add 50 µL of each radiolabeled polyamine solution to individual wells in a row, then work rapidly across the plate until all 60 wells have been treated (this step may be varied, *see* **Note 4**).
12. Start the timer (*see* **Note 8**).
13. Put a reagent trough in the ice tray and fill with PBS.
14. At the end of the incubation period immediately transfer the plate to melting ice.
15. Rapidly add 100 µL of ice-cold PBS to each well using a multichannel pipet.
16. Using the multichannel pipet set to 225 µL remove the medium from each well and discard (**Caution:** This waste is radioactive).
17. Add ~180 µL of ice-cold PBS to each well.
18. Remove the PBS and discard.
19. Repeat **steps 16** and **17** twice more to ensure cells are rinsed free from unbound extracellular polyamine (*see* **Note 9**).

At this point the plate can be can be stored at 4°C while other plates are processed.

3.4. Measurement of Cell Protein

The method described is that of Bradford ***(15)***, using a commercially prepared reagent.

1. Dissolve 10 mg bovine serum albumin (BSA) in 10 mL PBS, store at 4°C for up to 4 wk.
2. Dilute 100 µL of the 1 mg/mL BSA solution to 1 mL with PBS.
3. Add 5, 10, 15, 20, and 25 µL of the diluted BSA solution (containing 0.5, 1.0, 1.5, 2.0, and 2.5 µg of protein) to triplicate wells on the periphery of the 96-well plate, starting at well A2.
4. Dilute sufficient Protein Assay Dye Reagent Concentrate (Bio-Rad Laboratories GmbH, München, Germany) 1 to 10 with distilled water.
5. Add 100 µL of diluted dye reagent to wells containing cells, and standards, also to wells A1 to H1, which will act as blanks.
6. Leave the plates for 30 min at room temperature to allow the color to develop
7. Measure the absorbances at 620 nm in a microplate reader (e.g., Multiskan plate reader; Flow Laboratories, Irvine, Scotland).

If the plate reader is connected to a computer, the data can be saved to disk as a text file.

3.5. Measurement of Radiolabel Taken Up

1. Add 100 μL formic acid (25*M*) to the blue protein-dye solution in each cell-containing well and to six wells containing protein standards with absorbances close to those of the cell-containing wells; these will be used for standards and background. (**Caution:** *See* **Note 10**). The acid will digest all cellular material in the wells.
2. After 10 min transfer the contents of each well to a polyethylene mini β-vial.
3. Rinse each well by the addition of 200 μL water and transfer to the appropriate β-vial.
4. Add 10 μL of the 10 μ*M* radiolabeled polyamine solution to three of the β-vials containing protein standards, these will be used to determine the specific activity of the radiolabeled solutions used.
5. Dispense 3 mL of Scintran-T scintillation fluid from a container fitted with a bottle dispenser (e.g., Gilson Distrivar, Anachem Ltd. [Luton, UK]) into each vial.
6. Determine the amount of radioactivity per vial (as disintegrations per minute, dpm) using a liquid-scintillation counter. Count the three background vials (containing protein but no radiolabel) first.

If the counter is connected to a computer, the data can be saved to disk. Samples, background, and radiolabeled substrate are all counted under the same conditions of quench (sample color) and efficiency, thus reducing the errors and the need for extensive corrections.

3.6. Data Analysis

Rates of transport, expressed as picomoles of polyamine taken up per microgram of protein per hour, or in any other appropriate units, can be calculated by importing the data from the protein assay and the scintillation counter into a spreadsheet template. In the author's laboratory, values for the specific activity of the polyamine substrate are calculated for each experiment by dividing the mean dpm of the three standards by 100 (10 μL of a 10 μ*M* solution contain 100 pmol of polyamine) to give dpm/pmol. This value is then used to calculate uptake in pmol/well/h. Division by the protein content per well (in micrograms) gives uptake in pmol/μg protein/h. Apparent kinetic constants (K_m, sometimes written as K_t, and V_{max}) can then be estimated by computer-fitting of this data to Michaelis-Menten or other appropriate kinetic plots, using software that employs an adequate curve-fitting algorithm (e.g., Fig. P, Biosoft, Cambridge, UK, **Fig. 1**).

Nonspecific binding or noncarrier-mediated uptake can be estimated by carrying out the procedure in **Subheading 3.3.** at 0°C, with all solutions, and the 96-well plate, in melting ice (**Fig. 2**). The values so obtained can then be subtracted from those obtained at 37°C to derive true transport data. Kinetic constants for inhibition can be determined following preincubation of the cells with the inhibitor or concomitant addition of inhibitor and substrate.

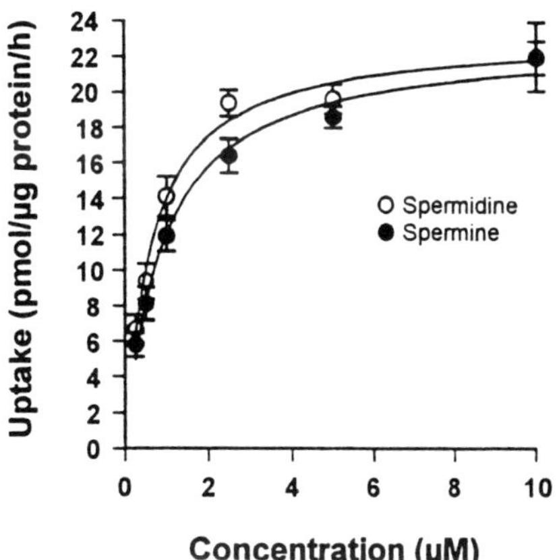

Fig. 1. Uptake of spermine (K_m 0.9 μM, V_{max} 22.9 pmol/µg protein/h) and spermidine (K_m 0.7 μM, V_{max} 23.2 pmol/µg protein/h) by ECV304 cells. Curves were plotted by fitting the data to the Michaelis-Menten equation after correction for nonspecific binding.

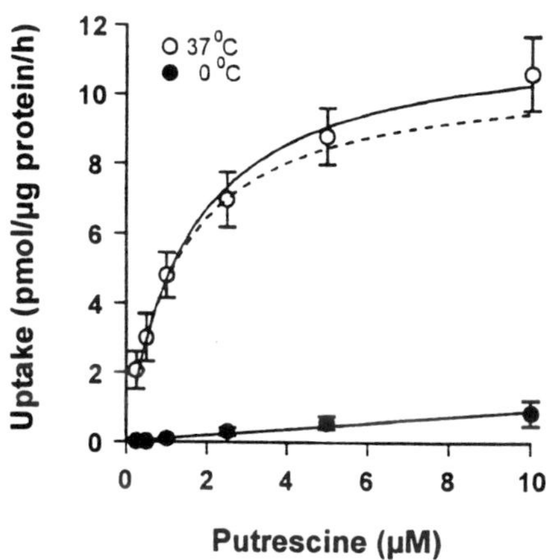

Fig. 2. Uptake of putrescine by ECV304 cells at 37 and 0°C. The dotted line, obtained by subtraction, shows a K_m of 1.5 μM and V_{max} of 11.8 pmol/µg protein/h.

4. Notes

1. Two different methods have been adopted for the use of radiolabeled tracers in transport studies. In the protocol described here, the specific activity is kept constant and the concentration of the radiolabel decreases with dilution of the substrate. This approach simplifies the subsequent calculations and is necessary where, as is the case with spermine, the quantity of radiolabeled species used

makes a significant contribution to the substrate concentration. The alternative is to maintain a constant level of radioactivity, independent of substrate concentration, but consequently with differing specific activities in each solution. Hence, the resulting calculations are more complex. This procedure can only be used when the addition of radiolabeled material makes an insignificant contribution to the total substrate concentration.

2. This firm was taken over by ICN Pharmaceuticals and unfortunately both these items and those mentioned in **Subheading 2., item 8** may no longer be available.
3. The main buffering system in HBSS is bicarbonate-CO_2, with a lesser contribution from phosphates. In order to maintain a steady pH in a low CO_2 environment, such as air, additional non-CO_2-dependent buffering capacity is required, which is provided by adding HEPES.
4. Multichannel repetitive dispensers are expensive. A single channel repetitive pipet can be used, working along one row at a time, starting with the lowest concentration and progressing upward. However, the timing becomes more complicated.
5. Transport must be measured at a point when the time-course for uptake is linear. In the case of ECV304 cells this so up to 20 min. For other cell types a suitable time-point must be found by preliminary experiments.
6. Calf serum contains an amine oxidase that will oxidize polyamines to cytotoxic aminoaldehydes ***(16)***. Failure to remove all serum-containing medium before addition of the polyamine solutions will result in low uptake because of partial metabolism of the polyamine and cell damage.
7. Care must be taken when inserting pipet tips into cell-containing wells not to damage the cell monolayer. It is a good practice to slide the pipet tip down one wall of the well until it reaches the junction with the base and subsequently always to use the same point of entry.
8. When a degree of proficiency has been obtained, if the incubation time is 15 min, then a second plate can be started while the first is incubating. Harvesting can then be done sequentially.
9. Addition of 1 m*M* unlabeled polyamine to the wash did not affect the results. The efficiency of the wash can be checked by including an extracellular marker, such as [^{14}C]mannitol, in the polyamine solution and counting both labels in the sample.
10. Formic acid is strongly corrosive. The vapor is lacrimatory and inhalation can result in irritation of the mucous membranes. In case of skin contact, rinse immediately with plenty of water.

Acknowledgment

This work was supported by the British Heart Foundation.

References

1. Seiler, N. and Dezeure, F. (1990) Polyamine transport in mammalian cells. *Int. J. Biochem.* **22,** 211–218.
2. Khan, N. A., Quemener, V., and Molinoux, J.-Ph. (1991) Polyamine membrane transport regulation. *Cell Biol. Int. Rep.* **15,** 9–24.

3. Pohjanpelto, P. (1976) Putrescine transport is greatly increased in human fibroblasts initiated to proliferate. *J. Cell Biol.* **68,** 512–520.
4. Feige, J. J. and Chambaz, E. M. (1985) Polyamine uptake by bovine adrenocortical cells. *Biochim. Biophys. Acta* **846,** 93–100.
5. Morgan, D. M. L. (1992) Uptake of polyamines by human endothelial cells: characterization and lack of effect of agonists of endothelial function. *Biochem. J.* **286,** 413–417.
6. Bogle, R. G., Mann, G. E., Pearson, J. D., and Morgan, D. M. L (1994) Endothelial polyamine uptake: selective stimulation by L-arginine deprivation or polyamine depletion. *Amer. J. Physiol.* **266,** C776–C783.
7. Alhonen-Hongisto, L., Seppänen, P., and Jänne, J. (1980) Intracellular putrescine deprivation induces uptake of the natural polyamines and methylglyoxal bis(guanylhydrazone). *Biochem. J.* **192,** 941–945.
8. Parys, J. B., De Smedt, H., Van Den Bosch, L., Geuns, J., and Borghgraeff, R. (1990) Regulation of the Na^+-dependent and the Na^+-independent polyamine transporters in renal epithelial cells (LLC-PK1). *J. Cell. Physiol.* **144,** 365–375.
9. Fouce, R. B., Escribano, M. I., and Alunda, J. M. (1991) Putrescine uptake regulation in response to α-difluoromethylornithine treatment in *Leishmania infantum* promastigotes. *Mol. Cell. Biochem.* **107,** 127–133.
10. Bey, P., Danzin, C., and Jung, M. (1987) Inhibition of basic amino acid decarboxylases involved in polyamine biosynthesis, in *Inhibition of Polyamine Metabolism* (McCann, P. P., Pegg, A. E., and Sjoerdsma, A., eds.), Academic, San Diego, CA, pp. 1–31.
11. Mitchell, J. L. A., Diveley, R. R., and Bareyal-Leyser, A. (1992) Feedback repression of polyamine uptake in mammalian cells requires active protein synthesis. *Biochem. Biophys. Res. Commun.* **186,** 81–88.
12. Mitchell, J. L. A., Judd, G. G., Bayeral-Leyser, A., and Luig, S. Y. (1994) Feedback repression of polyamine transport is mediated by antizyme in mammalian tissue culture cells. *Biochem. J.* **299,** 19–22.
13. Takahashi, K., Sawasaki, Y., Hata, J.-I., Mukami, K., and Goto, T. (1990) Spontaneous transformation and immortalization of human endothelial cells. *In Vitro Cell. Dev. Biol.* **25,** 265–274.
14. Morgan, D. M. L. (1996) Isolation and culture of human umbilical vein endothelial cells, in *Human Cell Culture Protocols* (Jones, G. E., ed.), Humana, Totowa, NJ, pp. 101–109.
15. Bradford, M. M. (1976) A rapid and sensitive method for the quantitation of microgram quantities of protein utilizing the principle of protein-dye binding. *Anal. Biochem.* **72,** 248–254.
16. Morgan, D. M. L. (1989) Polyamine oxidases and oxidised polyamines, in *The Physiology of Polyamines*, vol. 1 (Bachrach, U. and Heimer, Y. M., eds.), CRC, Boca Raton, FL, pp. 203–229.

18

Measurement of Polyamine Transport

Cells in Suspension

Sarah A. Le Quesne and Alan H. Fairlamb

1. Introduction

Most cells are able to supplement *de novo* polyamine synthesis with uptake on specific transport systems, especially in response to stimuli that provoke cell growth or when *de novo* synthesis is blocked by specific inhibitors, such as D,L-α-difluoromethylornithine *(1,2)*.

The method described here is a rapid sampling technique, which involves the separation of cells from radiolabel by sedimentation through silicone oil. It was originally used for the measurement of nucleoside transport in lymphoma cells *(3)*. Subsequently it has been used in parasitic protozoa to measure nucleoside *(4)*, drug *(5)*, and polyamine transport *(6)*. However, this method can be adapted for use with any radiolabeled compound or cell type that grows in suspension, provided the radiolabel is not miscible with the oil and the cells can be rapidly sedimented by centrifugation through the oil layer. Although technically demanding and requiring a bit of initial effort to establish optimal assay conditions, the advantages of the method are well worth the investment. Transport measurements can be made over very short time intervals (for example, 1–5 s), thereby reducing the possibility of metabolism or excretion of the radiolabel following uptake. Many (100+) samples can be processed in an afternoon, thereby enabling a large quantity of information to be obtained relatively quickly. Removal of the cells from the radiolabel by centrifugation through oil negates the need to stop the reaction by adding excess cold polyamine or transport inhibitors. The precision of the initial rates of uptake tends to be better than other methods, since for any single substrate or inhibitor concentration, the rate has been determined by linear regression on four to five data points measured at several time intervals. Finally, using this approach

From: *Methods in Molecular Biology, Vol. 79: Polyamine Protocols*
Edited by: D. Morgan Humana Press Inc., Totowa, NJ

there is no requirement to correct for binding or carry-over of radiolabel with the cells.

2. Materials

1. Cells: The experimental procedure described here has been used successfully to measure polyamine transport of several parasitic protozoa, namely *Trypanosoma cruzi* (line MHOM/BR/78/Silvio, clone X10/6) *(6)*, *Trypanosoma brucei*, *Leishmania donovani*, and *Crithidia fasciculata* (*see* **Note 1**). However, because this method has also been used to measure the uptake of other radiolabeled compounds in mammalian cells *(3)*, it should be possible to adapt the procedure to measure polyamine transport in any cell type that can grow in suspension.
2. CBSS, pH 7.4 *(7)*: This balanced salt solution essentially consists of the inorganic salts, *N*-[2-hydroxyethyl]piperazine-*N*'-[2-ethanesulfonic acid] (HEPES), and glucose present in RPMI 1640 medium. To prepare, dissolve 25 mmol HEPES, 120 mmol NaCl, 5.4 mmol KCl, 0.55 mmol $CaCl_2$, 0.4 mmol $MgSO_4$, 5.6 mmol Na_2HPO_4, and 11.1 mmol D-glucose in 800 mL of double-distilled water (ddH_2O). Add 1 mL of phenol red (5 mg/mL). Adjust to pH 7.4 with 1*M* NaOH and make up to 1 L with ddH_2O. Store at 4°C for up to 2 wk or 1–2 mo if it has been filter sterilized (*see* **Note 2**).
3. Silicone oil: Tetrachlorophenyl-modified silicone oil is used (Versilube F-50; viscosity 75 centistokes; specific gravity 1.05 g/mL, Medford Silicones, NJ) (*see* **Note 3**).
4. Radiolabeled polyamines: [1,4<n>-^{3}H] putrescine. 2HCl (407 GBq/mmol, Amersham International plc), [terminal methylenes ^{3}H(N)-] spermidine. 3HCl (577 GBq/mmol, New England Nuclear), or [terminal methylenes ^{3}H(N)-] spermine. 4HCl (1110–2220 GBq/mmol, American Radiolabeled Chemicals) (*see* **Note 4**). Store at –20°C. Stable for several mo (about 1–5% decomposition/yr) (*see* **Note 5**).
5. Phosphate-buffered saline (PBS): 0.01*M* sodium phosphate, pH 7.4, and 0.145*M* NaCl. Make up as a 10X stock containing 84.5 m*M* Na_2HPO_4, 15.5 m*M* NaH_2PO_4, and 1.45*M* NaCl. Dilute 1:10 in ddH_2O before use (check that the pH is 7.4). Can be kept for about 1 mo at room temperature.
6. 1*M* NaOH.
7. Scintillation fluid: Pico-Fluor 40, Canberra Packard Ltd. (*see* **Note 6**).
8. Microfuges. We found that the Eppendorf (5415C) microfuge fitted with a fixed angle rotor worked best because it had a faster acceleration time than the other microfuges that we tested, and hence the cells showed faster sedimentation through the oil layer.

   If the cells are to be sedimented through silicone oil into trichloroacetic acid (*see* **Subheading 3.5.**) then a microfuge fitted with a horizontal bowl rotor is recommended (Microfuge E or GS15, Beckman). Fixed-angle rotors, even if fitted with adaptors to take 0.4 mL tubes, do not work well because the cells are preferentially sedimented against the side wall of the tube rather than through the oil layer.
9. A Low vacuum diaphragm pump (CAPEX 2DC, Charles Austen pumps, supplied by Merck Ltd.) is used for aspiration of the radiolabel and oil. It is attached by

silicone tubing to a fine-tip glass Pasteur pipet via a conical flask trap (containing 20% Chloros if you are working with pathogenic organisms).

3. Methods

3.1. Harvesting the Cells

1. Pellet the cells in the exponential phase of growth by centrifugation (1500*g*, 10 min, 4°C). Remove the supernatant and gently resuspend in CBSS using a standard-bore Pasteur pipet.
2. Repeat the aforementioned centrifugation procedure two times, only resuspend the final pellet to 2×10^8 cells/mL in CBSS. Store the prepared cells on ice; they must be used within 1.5 h of preparation.

3.2. Pelleting Silicone Cells Through Oil

1. Into 1.5 mL microfuge tubes pipet 100 µL of silicone oil (*see* **Note 7**). Use five tubes for each substrate (inhibitor) concentration. Overlay with 100 µL of CBSS containing radiolabeled polyamine ([^{3}H]putrescine, [^{3}H]spermidine, or [^{3}H]spermine), unlabeled polyamine, and any inhibitor at two times the desired final concentration (*see* **Note 8**). Centrifuge the tubes briefly in a microfuge (16,000*g*, 15 s) to ensure separation of the silicone oil and label into two layers.
2. Prewarm a suitable aliquot of cells (for example, 0.8 mL) and five Eppendorf tubes containing the radiolabel to 28°C (or appropriate temperature) for 10 min in a heating block.
3. Quickly transfer the microfuge tubes (caps facing outward) to a microfuge fitted with a fixed angle rotor. At time intervals of 5 s (*see* **Note 9**), with one hand sequentially add 100 µL aliquots of cells to each of the tubes by pipeting down the side of the tube wall. This ensures rapid and adequate mixing. With the other hand, cap the tubes after the addition of each sample. Five seconds after the addition of the last sample, separate the cells from the medium by centrifugation (16,000*g*, 1 min) (**Fig. 1**).
4. Store samples on ice for subsequent processing (*see* **Note 10**).
5. Repeat the process with the next set of tubes containing a different substrate concentration.

3.3. Extraction and Counting of Cell Pellets

1. When all the samples are completed, aspirate the medium using a vacuum pump fitted with a trap. Rinse the region above the oil layer with 0.8 mL PBS by carefully adding the PBS from a pipet in a circular motion around the top of the tube; then aspirate the PBS layer. Repeat this step once more. This removes any residual label prior to aspirating the silicone oil by placing the Pasteur pipet (which is attached to the vacuum pump) down the wall of the microfuge tube on the opposite side to the cell pellet, taking great care not to dislodge it (**Fig. 1**).
2. Extract the polyamine from the washed pellet overnight with 0.1 mL 1*M* NaOH (*see* **Note 11**). Add 1 mL of scintillant (Pico-fluor 40) to the microfuge tubes,

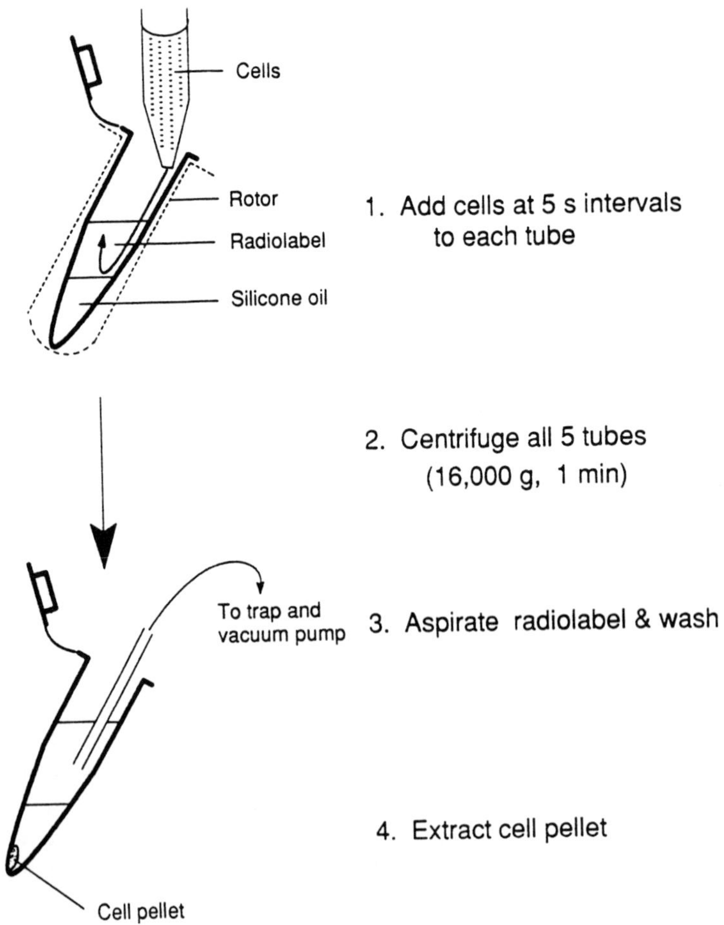

Fig. 1. Principle of rapid transport method.

cap, and gently vortex to mix the contents. Transfer the microfuge tubes to special inserts or carrier vials that then fit into the scintillation counter racks and determine the amount of radioactivity using a suitable liquid scintillation counter (Beckman LS 6000LL).

3.4. Analysis

3.4.1. Initial Rates

Initial rates of uptake (dpm/s) are determined by linear regression analysis on at least four time-points spaced at 5 s intervals over the first 20–25 s. This approach obviates the need to correct for nonspecific binding to cells, carry-over of extracellular fluid through the silicone oil, or the time lag from starting the centrifuge until the cells have entered the oil, since the intercepts are not used in subsequent calculations. All rates should have a regression coefficient of $r > 0.95$. Less than 10% of the available radiolabel should be taken up over the whole time-course, otherwise the substrate concentration cannot be regarded as constant through out the whole experiment.

If you are having trouble in obtaining linear rates, some points to look out for are given below.

1. Low counts:
   a. Insufficient label being taken up over the time-course. Try incubating the cells for longer time intervals or increase the specific activity of the [^{3}H]polyamine in the assay mixture.
   b. The cell pellet is being dislodged when aspirating the liquid. This can also cause erratic counts.
   c. On storage on ice prior to aspiration, there may be loss of radiolabel from the cell pellet into the silicone oil. Check by storing the cells on ice for different lengths of time before aspirating the radiolabel. If there is no loss of label from the cell pellet then the counts should be constant with time.
2. High counts:
   a. Insufficient washing (removal) of unincorporated radiolabel or residual binding of the radiolabel to the tube walls. Check by scintillation counting a tube containing only radiolabel and oil that has undergone the same centrifugation and extraction steps as the other tubes. If binding to the tube is a problem (possible with spermine) then siliconize the tubes first using a commercially available siliconizing agent (Merck Ltd. or Sigma).
   b. If all of your tubes have the same high level of counts in them, then all the label may have been taken up by the first point in the time-course. Try measuring at shorter time intervals, using fewer cells or at a lower temperature.

3.4.2. Calculations

1. Determine the initial rate of uptake by unweighted linear regression (dpm/s).
2. Convert into nmol/min/(10^8 cells) using the following formula:

$$v\,(6 \times 10^9)/(a\,n) \tag{1}$$

   where:

   v = initial rate (dpm/s).
   a = final specific activity of radiolabel in incubation mixture (dpm/nmol).
   n = number of cells added to each microfuge tube.
3. If uptake obeys simple Michaelis-Menten type kinetics then affinity of the substrate for the transporter (K_m) and maximum rate of transport (V_{max}) can be calculated graphically by various linear transformations ***(8)***. However, we recommend fitting the data to the Michaelis-Menten equation by unweighted nonlinear regression using an appropriate curve-fitting program, such as Enzfitter (Elsevier Biosoft, Cambridge, UK) or GraFit (Erithacus Software Limited, Staines, UK).

For more complex situations, e.g., uptake by more than one transporter or transport plus diffusion, the reader should consult **refs.** ***9*** and ***10***.

3.5. Pelleting Cells Through Silicone Oil into Trichloroacetic Acid

This method is an adaption of that described in **Subheading 3.2.** It can be used to investigate both the uptake and subsequent metabolism of the radiolabeled polyamine. Metabolism is arrested by centrifuging the cells through silicone oil into 10% trichloroacetic acid.

1. Prepare cells as described in **Subheading 3.1.**
2. Prepare 0.4 mL polyethylene tubes (*see* **Note 12**), by addition of 100 µL of 10% trichloroacetic acid made up in 10 m*M* HCl. Overlay with 100 µL of silicone oil. Centrifuge (16,000*g*, 15 s) to ensure separation of the oil and trichloroacetic acid into two layers.
3. Prewarm cells and radiolabel to 28°C (or appropriate temperature).
4. At time zero add an equal volume of cells to the radiolabel (both made up at twofold final concentration) in a microfuge tube. Mix and immediately withdraw the mixture with an Eppendorf Multipet 4780. Pipet 100 µL aliquots on top of the oil layer in the prepared 0.4-mL tubes.
5. Cap the tubes and place in the fixed horizontal rotor of a microfuge. At regular time intervals centrifuge for 1 min (12,500*g*) to sediment the cells through the oil (*see* **Note 13**).
6. Leave the tubes overnight at 4°C for full extraction of the polyamines. We found two suitable methods for obtaining the trichloroacetic acid layer without disturbing the upper layers. The tube can be cut just at the interface between the acid extract and the silicone oil using either a tube cutter or a scalpel. Or, with the caps open, pierce the side of the tube wall, at the bottom of the trichloroacetic acid layer, with a fine-bore needle (for example, 26 gage, 0.5 in.) and withdraw the sample into a 1 mL syringe. Transfer an aliquot for scintillation counting and extract the remainder five times with ethyl acetate, dry down, derivatize with dansyl chloride, and analyze by HPLC for polyamine content ***(11)*** (*see also* Chapter 14).

4. Notes

1. All operations with *T. cruzi* and *L. donovani* must be carried out in a Class II safety cabinet.
2. Any basic salts solution that will keep the cells alive and metabolically active for the duration of the uptake experiments can be used.
3. Before starting out, it is important to establish that your cells will sediment through the silicone oil to form a visible cell pellet. Ideally >95% should pass through the oil within 10 s of spinning. This can be checked by counting the number of cells remaining in the layer above the oil at different times. If separation is not satisfactory, this may be a result of the density and/or viscosity of the silicone oil. Silicone oils of different densities can be prepared by mixing Dow

Corning 550 fluid and Dow Corning 510 fluid, 50 centistokes ***(12)*** or Aldrich silicone oil and paraffin oil ***(3)***. Dibutylphthalate/dinonylphthalate have also been used ***(13)***. Formulation of the optimal mixture can be determined empirically. Alternatively, the concentration of the cells in the incubation medium may be too low. For *T. cruzi* a minimum of 1×10^7 cells per tube is required.

4. Commercially available [^{14}C]spermidine or [^{14}C]spermine are less suitable because of their 1000-fold lower specific activity than the equivalent tritiated ones. (For those unfamiliar with S.I. units of radioactivity 1 µCi = 37 kBq, where 1 Bq = 1 disintegration/s).
5. Aliquots of the radiolabel can be analyzed for purity by HPLC. If necessary, the radiolabel can be repurified by separation of the [^{3}H]polyamine from impurities by HPLC with the postcolumn derivatization turned off (*see* Chapter 15). Collect the [^{3}H]polyamine peak and dilute with 2 vol of ddH$_2$O. Connect a C$_{18}$-Sep-Pak cartridge (Waters, Millipore Ltd.) to a 5 mL plastic syringe and wash the resin with 2 mL of propan-1-ol followed by 5 mL of ddH$_2$O. Apply the pooled sample to the column using a syringe, wash with 5 mL ddH$_2$O to remove the HPLC salts and ion-pairing reagents, and elute with 5 mL of propan-1-ol. Dry the propan-1-ol fraction by centrifugal evaporation under reduced pressure and resuspend in a small volume of 10 m*M* HCl.
6. Use any scintillation cocktail that can tolerate at least 20% by volume aqueous (alkaline, 1*M* NaOH) solution as a single phase and does not suffer from excessive quench or chemiluminescence.
7. Use of an Eppendorf Multipet 4780 fitted with a 5 mL Combitip (volume selector: 1 = 100 µL) greatly increases the speed with which oil, label, 1*M* NaOH, and cells (for these use a 1.25 mL Combitip fitted with a yellow 2–200 µL pipet tip) can be added to the tubes.
8. This is most easily achieved by making up the radiolabel in the highest substrate concentration that you are using (i.e., 50–100 µ*M*) and then serially diluting this stock solution (*see* also Note 1, Chapter 17).
9. For time intervals of 5 s and over an electronic timer is adequate. If time intervals of <5 s are required, use of a metronome is recommended to ensure accurate timings. In this case microfuge tubes with the caps cut off must be used to avoid delays caused by manipulating the cap.
10. *T. cruzi* can be stored for at least 30 min on ice with no loss of radiolabel from the pellet, but this should be checked for your own type of cells.
11. The 1*M* NaOH can be replaced with detergents (e.g., 1% Triton X-100) or other denaturing agents, such as 10% trichloroacetic acid and 12% perchloric acid.
12. Use the 0.4 mL polyethylene Eppendorf tubes (Merck Ltd.) as these work well, whereas the tubes from the other suppliers that we tested had a tendency to split.
13. If measuring time intervals of <90 s, this requires the simultaneous use of several microfuges. With an assistant, duplicate samples can be centrifuged every 15 s.

Acknowledgment

The authors would like to thank the Wellcome Trust for their financial assistance.

References

1. Seiler, N. and Dezeure, F. (1990) Polyamine transport in mammalian cells. *Int. J. Biochem.* **22,** 211–218.
2. Khan, N. A., Quemener, V., and Moulinoux, J.-P. (1994) Characterization of polyamine transport pathways, in *Neuropharmacology of Polyamines* (Carter, C., ed.) Academic, Ltd. London, pp. 37–60.
3. Aronow, B., Allen, K., Patrick, J., and Ullman, B. (1985) Altered nucleoside transporters in mammalian cells selected for resistance to the physiological effects of inhibitors of nucleoside transport. *J. Biol. Chem.* **260,** 6226–6233.
4. Carter, N. S. and Fairlamb, A. H. (1993) Arsenical-resistant trypanosomes lack an unusual adenosine transporter. *Nature* **361,** 173–176.
5. Carter, N. S., Berger, B. J., and Fairlamb, A. H. (1995) Uptake of diamidine drugs by the P2 nucleoside transporter in melarsen-sensitive and -resistant *Trypanosoma brucei brucei. J. Biol. Chem.* **270,** 28,153–28,157.
6. Le Quesne, S. A. and Fairlamb, A. H. (1996) Regulation of a high affinity diamine transport system in *Trypanosoma cruzi* epimastigotes. *Biochem. J.*, **316,** 481–486.
7. Fairlamb, A. H., Carter, N. S., Cunningham, M., and Smith, K. (1992) Characterisation of melarsen-resistant *Trypanosoma brucei brucei* with respect to cross-resistance to other drugs and trypanothione metabolism. *Mol. Biochem. Parasitol.* **53,** 213–222.
8. Segel, I. H. (1975) *Biochemical Calculations.* 2nd ed., Wiley, New York.
9. Stein, W. D. (1989) Kinetics of transport: analyzing, testing, and characterizing models using kinetic approaches. *Methods Enzymol.* **171,** 23–62.
10. Stein, W. D. (1990) *Channels, Carriers and Pumps: An Introduction to Membrane Transport.* Academic, London.
11. Hunter, K. J., Le Quesne, S. A., and Fairlamb, A. H. (1994) Identification and biosynthesis of N^1,N^9-bis(glutathionyl)aminopropylcadaverine (homotrypanothione) in *Trypanosoma cruzi. Eur. J. Biochem.* **226,** 1019–1027.
12. Strauss, P. R., Sheehan, J. M., and Kashket, E. R. (1976) Membrane transport by murine lymphocytes. I. A rapid sampling technique as applied to the adenosine and thymidine systems. *J. Exp. Med.* **144,** 1009–1021.
13. Fasoli, M. O. and Kerridge, D. (1990) Uptake of pyrimidines and their derivatives into *Candida glabrata* and *Candida albicans. J. Gen. Microbiol.* **136,** 1475–1481.

19

Measurement of Polyamine Efflux from Cells in Culture

Heather M. Wallace and A. Jill Mackarel

1. Introduction

The polyamines are cellular growth factors in most living organisms. Cells that are rapidly growing require high polyamine concentrations, whereas quiescent cells maintain comparatively low polyamine concentrations *(1,2)*. This was shown to be the case in normal baby hamster kidney (BHK) fibroblasts grown to confluence *(3)*. Here cells that were undergoing density-dependent inhibition of growth had significantly lower total polyamine content than cells in exponential growth. The question was, then, how did the cell decrease its polyamine content in response to the decreased growth rate? Analysis of the fate of radiolabeled polyamines showed that polyamines were released from quiescent cells into the extracellular medium *(4–6)*. The efflux was found to be specific for N^1-acetylspermidine and to be regulated in concert with the growth rate of the cell *(3,7)*. Efflux is an integral part of the regulatory process responsible for controlling the intracellular polyamine content of cells. In addition, such factors as drugs *(8,9)*, virus infection *(10,11)*, and cell transformation *(12)* have all been found to alter the efflux of polyamines from cells in culture.

Because the predominant efflux product is N^1-acetylspermidine and the major intracellular polyamine in most cell types is spermine, metabolism of the higher polyamines is necessary. Metabolism involves two enzyme reactions, the N^1-acetyltransferase and the polyamine oxidase, as outlined in **Fig. 1.** The rate-limiting reaction for this pathway is the acetyltransferase, which is itself a highly inducible, rapidly turned over enzyme (for review *see* **ref. *13*** and for measurement *see* Chapter 6 of this volume). Thus, efflux is an integral part of the regulatory process responsible for controlling the intracellular polyamine content of cells.

From: *Methods in Molecular Biology, Vol. 79: Polyamine Protocols*
Edited by: D. Morgan Humana Press Inc., Totowa, NJ

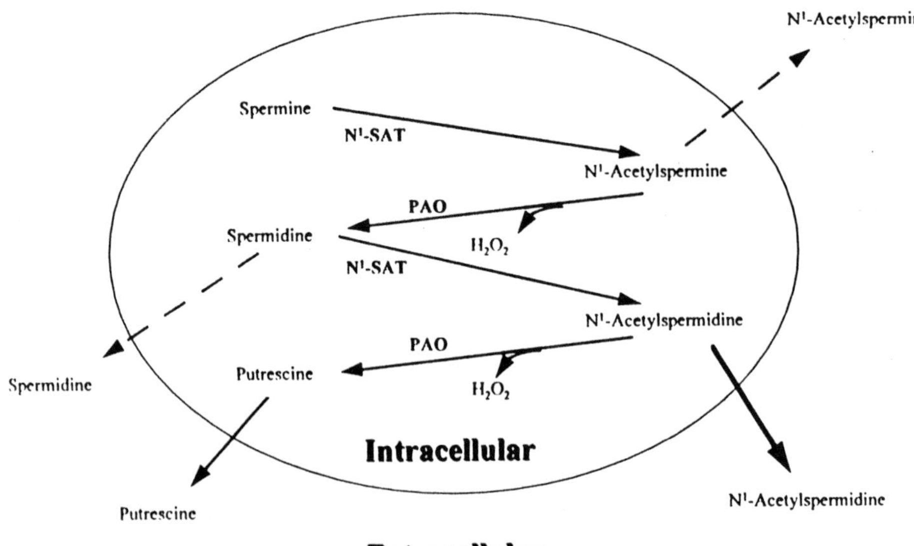

Fig. 1. Metabolism and efflux of polyamines from mammalian cells. Efflux products: — major, — minor, --- possible.

In humans, polyamines are excreted in the urine mainly in the form of monoacetylated derivatives ***(14)***, and this excretion is increased in disease states, such as cancer ***(15)*** and cystic fibrosis ***(16)***, and in normal physiological conditions, such as pregnancy ***(17)***. In the early 1970s it was thought that urinary polyamine measurements could be used in the diagnosis of malignant disease. However, a number of false positive results showed this was not a useful biological marker of the presence of neoplasia. Measurement of urinary polyamine content does, however, show some promise as a means of monitoring the response of cancer patients to therapy with patients in remission having urinary polyamine concentrations within the normal range ***(18)***.

1.1. Principle

Intracellular polyamine pools are radiolabeled from the precursor, [^{3}H] putrescine. Radioactive and free polyamine pools are allowed to equilibrate and then intracellular and extracellular polyamine content is monitored. Individual polyamines are separated and identified by HPLC using both fluorescent and radiomatic detectors.

1.2. Background

It is important to label all the polyamine pools, so a minimum labeling time of one generation time is required. Because polyamines are positively charged they will bind electrostatically to any negatively charged surface e.g., the cell surface, so it is essential to wash the cells thoroughly to remove any nonspecifically bound polyamines from the extracellular surface before beginning the experiment. As much as 30% of the total polyamine associated with the cell at the end of the initial labeling period can be nonspecifically bound. Serum, especially ruminant serum, is known to contain significant quantities of copper amine oxidase, which is capable of metabolizing both spermidine and spermine to toxic aldehydes ***(19,20)***, therefore, it is necessary to either include 1 m*M* aminoguanidine, a copper amine oxidase inhibitor, in the medium or use nonruminant serum with low amine oxidase activity, such as horse serum. We have used the latter strategy successfully and have found little metabolism of spermine over 48 h in culture medium containing horse serum.

2. Materials

2.1. Cell Culture

1. Sterile 75-cm^2 tissue culture flasks.
2. Sterile tissue-culture plates (e.g., 5 cm^2 Nunclon plates, Life Technologies, Paisley, UK).
3. 25-mL sterile Universal containers.
4. 10- and 25-mL sterile plastic pipets.
5. Sterile 0.2-µm filter units.
6. Sterile 100- and 500-mL glass bottles.
7. Dulbecco's Modification of Eagle's medium (DMEM): purchased in powder form without pyruvate or bicarbonate (136.5 g/tub; 1 tub makes up 10 L culture medium, *see* **step 11**) and stored at 4°C.
8. Horse serum (mycoplasma screened): stored as 50-mL aliquots in sterile 100-mL glass bottles at –20°C for up to 6 mo.
9. Penicillin-streptomycin solution (purchased as 100-mL sterile solution of 5000 IU/mL penicillin, 5000 µg/mL streptomycin, Gibco-BRL, Life Technologies, Paisley, UK).
10. HT115 cells (*see* **Note 1**; Human colorectal carcinoma cell line, European Collection of Animal Cell Cultures, Porton Down, UK).
11. Culture medium: Dissolve 1 tub of DMEM (136.5 g), 1.1 g Na pyruvate, and 37 g $NaHCO_3$ in 5 L of nonpyrogenic, deionized sterile water. Add 100 mL penicillin-streptomycin solution and make up to 10 L, adjusting the pH to 7.2 using 1*M* NaOH. Sterilize the medium by membrane filtration through a 0.2 µ*M* pore-size filter (e.g., Gelman Science Microcapsule) and aliquot into 450 mL fractions. Store at 4°C in the dark.
12. Serum supplemented culture medium (DH_{10}): Defrost a 50 mL aliquot of horse serum and add to 450 mL DMEM, giving a final serum content of 10%. Store in the dark at 4°C and use within 2 wk.

2.2. Cell Harvesting

1. Phosphate-buffered saline (PBS): mix solutions A, B, and C in a volume ratio of 4:5:5:
   PBS A: 10% (w/v) NaCl, 1.75% (w/v) $Na_2HPO_4 \cdot 2H_2O$, 0.3% (w/v) KCl, 0.3% (w/v) KH_2PO_4, adjust pH to 7.2.
   PBS B: 0.1% (w/v) $CaCl_2 \cdot 2H_2O$.
   PBS C: 0.1% (w/v) $MgCl_2 \cdot 6H_2O$.
   Store PBS at 4°C.
2. Tris buffered saline (TBS): Make up a solution of 0.8% (w/v) NaCl, 0.01% (w/v) Na_2HPO_4, 0.1% (w/v) glucose, 0.3% (w/v) Trizma base, 0.04% (w/v) KCl, and 0.0015% (w/v) phenol red indicator in distilled, deionized water, adjusting the pH to 7.4 with HCl. Sterilize by autoclaving at 121°C for 20 min.
3. Trypan blue solution: add 0.1 g Trypan blue to 100 mL PBS A solution (*see* **Subheading 2.2.**, **step 1**) and leave stirring overnight. Filter through Whatman No. 1 filter paper and store at 4°C.
4. Trypsin solution: Dilute 2.5% sterile trypsin solution (purchased from Gibco-BRL, Life Technologies) with TBS (*see* **step 2**) to give a 0.25% solution. Dispense into 50-mL aliquots and store at –20°C.
5. Versene solution: Make up 0.5 m*M* EDTA (disodium salt) in PBS A containing 0.002% (w/v) phenol red indicator. Adjust pH to 7.4. Sterilize by autoclaving at 121°C for 20 min. Store at room temperature.
6. Trypsin/versene (T/V) solution: Add 50 mL trypsin solution to 100 mL versene solution. Store at 4°C.

2.3. Separation of Polyamines-HPLC

1. HLPC buffers:
   Buffer A: 0.1*M* Na acetate (adjust pH to 4.5 with glacial acetic acid) and 0.01*M* 1-octane sulfonic acid.
   Buffer B: 0.2*M* Na acetate (adjust pH to 4.5 with glacial acetic acid), 0.01*M* 1-octane sulfonic acid, and 23% (v/v) acetonitrile.
   Buffer C: Per liter of distilled water dissolve the following: 50 g boric acid, 44 g potassium hydroxide, 3.5 g Brij-35 solution, 0.4 g 2-phthalaldehyde in 5 mL methanol, and 2 mL 2-mercaptoethanol.
   Vacuum filter all buffers through 0.45-μm filters (Millipore) and degas with helium before use.
2. Polyamine external reference standard solution: Make up a stock solution of 2.5 m*M* putrescine and *N*-acetylputrescine, and 5 m*M* spermidine, spermine, N^1-acetylspermidine, N^8-acetylspermidine, and N^1-acetylspermine, in 0.2*M* perchloric acid. Store in 100-μL aliquots at –80°C. To prepare a reference standard, defrost an aliquot of stock solution and dilute 1 in 800 with 0.2*M* perchloric acid: 100 μL of this standard contains 0.312 nmol of putrescine and *N*-acetylputrescine, and 0.625 nmol of each of the other polyamines.
3. HPLC system:
   a. A reverse-phase μ Bondapack C18 (3.9 × 300 mm 10 μm particle size) HPLC column.

b. Two pumps (Waters, model 501) with an inbuilt manometric module.
c. A waters Intelligent Sample Processor (WISP 712) autoinjector.
d. An in-line precolumn filter unit.
e. Postcolumn Waters pump (model 6000a solvent delivery system).

Fluorescence spectrophotometer (Waters, 420-AC); fluorescence excitation was achieved at 347 nm and emission was measured at 465 nm.

Flo-1β radiomatic detector (Canberra Packard, Pangbourne, UK).

3. Methods

3.1. Harvesting of HT115 Cells

1. Warm T/V solution and DH_{10} to 37°C.
2. Remove confluent cultures of HT115 cells (growing in sterile plastic 75-cm^2 flasks) from the incubator. Pour off the medium. Gently rinse the cell sheet with 2–3 mL T/V to remove any traces of culture medium and discard. Add sufficient T/V solution to cover the surface of the culture flask and incubate the flask at 37°C for approx 5 min until the cells detach (*see* **Note 2**).
3. Once the cells have detached, add an equal volume of DH_{10} to the T/V-cell suspension to inactivate the trypsin.
4. Decant the cell suspension into sterile Universal containers and centrifuge at 800*g* for 3–4 min in a benchtop centrifuge to pellet the cells.
5. Carefully decant the supernatant. Resuspend and pool the resulting cell pellet in a total volume of 10 mL DH_{10}.
6. Transfer a 100-μL aliquot of the cell suspension into an Eppendorf, add 900 μL of Trypan blue solution, and gently mix by inverting. Count the cell number and viability using an Improved Neubauer hemocytometer (Weber Scientific International Ltd). Using the Trypan blue exclusion method, live cells remain colorless, but dead cells take up the dye and appear blue.

3.2. Radiolabeling Cellular Polyamines

1. Prepare a suspension of HT115 cells in DH_{10} medium as described in **Subheading 3.1.** Seed sterile, plastic tissue-culture plates with 2.8×10^4 cells/cm^2 of culture plate area in a volume of DH_{10} to give approx 1 mL medium/cm diameter of the culture plate.
2. Incubate cultures for 16 h (attachment period).
3. Add [^{3}H] putrescine (0.5 μCi/mL) and incubate cultures for 36 h (*see* **Note 3**).
4. After 36 h growth in the presence of [^{3}H] putrescine, carefully remove the culture medium and gently wash the cell sheet twice with 2–3 mL of DMEM (warmed to 37°C). This procedure removes any radiolabel nonspecifically bound to the cell surface (*see* **Note 4**).
5. Add fresh DH_{10}, incubate the cultures for a further 12 h, and then change the culture medium again. This is taken as time 0 h ($t = 0$). Efflux of cellular radioactivity is monitored from this time-point (*see* **Note 5**).

3.3. Monitoring Efflux of [^{3}H]-Labeled Polyamines

1. At selected time-points, remove cultures from the incubator and place on ice (4°C).
2. Pipet the medium into plastic 10 mL conical tubes, recording the volume, and centrifuge at 800*g* for 10 min to sediment any suspended cellular material. Carefully decant the medium supernatant into a clean conical tube at 4°C. Transfer a 1.0-mL aliquot of medium into an Eppendorf tube and add 0.25 mL of 1*M*-$HClO_4$. Invert the tube to mix and incubate at 4°C for 30 min to extract polyamines and precipitate protein. After the 30 min incubation time, centrifuge the tube at 13,000*g* for 3 min. Remove the acid-supernatant containing the extracted polyamines and store at –20°C until further analysis.
3. Having removed the culture medium, gently rinse the cell sheet three times with ice-cold PBS. Add 0.5 mL of 0.2*M* $HClO_4$ and incubate for 20 min at 4°C to precipitate cellular protein and extract polyamines. Scrape the protein from the surface of the culture plate using a rubber policeman or cell scraper and transfer into an Eppendorf tube. Rinse the culture plate with a further 0.25–0.5 mL of 0.2*M*-$HClO_4$ to remove any remaining precipitated protein and add this to the Eppendorf. Leave the protein suspension at 4°C for a further 5–10 min and then centrifuge at 13,000*g* for 3 min to sediment the protein fraction. Remove the polyamine-containing acid supernatant and store at –20°C until further analysis. Dissolve the pellet in 1.0 mL of 0.3*M* NaOH at 37°C and then assay for protein content (e.g., using the method of Lowry et al., ***21***).
4. Identification and quantification of effluxed [^{3}H]-labeled polyamines was performed using HPLC ***(22,23)*** followed by radiomatic detection: Defrost stored acid-extracted polyamine samples and then centrifuge at 13,000*g* for 5 min to pellet any suspended protein. Remove the supernatant and add 50 µL of cell extract or 450 µL of medium extract to a scintillation vial. Add 450 µL of cold, nonradioactive medium to the cell extract and 50 µL of 0.2*M* $HClO_4$ to the medium extract. Finally, add 4 mL of Ultima Gold (or other cocktail suitable for aqueous counting) scintillation cocktail to each vial, mix well, and determine the radioactivity in each fraction using a standard scintillation spectrophotometer with a tritium protocol. Addition of the radioactivity from both the cell and medium fractions gives the total radioactivity per culture. Intra- and extracellular radioactivity can then be expressed as a percentage of the total or can be related to the protein content of the culture.

4. Notes

1. Although this protocol outlines the procedure for monitoring polyamine efflux from HT115 cells, it has also been used with other anchorage-dependent cell lines, including BHK-21/C13 (baby hamster kidney) cells, T47-D (human breast carcinoma) cells, and the anchorage-independent T-cell line, MOLT-4. It could be adapted for use with other cell types, as discussed in **Note 2.**
2. This procedure for monitoring polyamine efflux can be adapted for use with other types of cell lines. It is important that the cells are seeded at a density that ensures they are in log phase of the growth cycle during the incubation with [^{3}H]-putrescine

Table 1
Comparison of the Distribution of [^{3}H]-Labeled Polyamines and the Total Cellular Polyamine Distribution of HT115 Cells of an Efflux Experiment[a]

	Cellular polyamine distribution, % of total[b]	
Polyamine	Total polyamines	Radiolabeled polyamines
Putrescine	9.09 ± 1.82	1.61 ± 0.25
Spermidine	37.14 ± 1.48	35.16 ± 2.94
Spermine	50.98 ± 1.39	59.63 ± 1.39
N^1-Acetylspermidine	2.80 ± 1.11	3.59 ± 0.47

[a]t = 0 h; n = 4.
[b]Values are mean ± SD. n = 4.

because extracellular polyamines are accumulated more readily by actively growing cells *(24)*. In log phase, the culture is in its most reproducible form with a high growth fraction of usually 90–100% of the culture actively proliferating *(25)*. Adding radiolabeled polyamine precursors during this phase of the growth cycle maximizes its cellular accumulation. For HT115 cells, we found a seeding density of 2.8 × 10^4 cells/cm^2 and an attachment period of 16 h ensured that any lag phase of growth was over before the polyamine precursor, [^{3}H]-putrescine, was added. The cells were in the log phase of the cell cycle during the 36 h incubation time with [^{3}H] putrescine and readily accumulated the polyamine.

3. The length of the radiolabeling period was chosen to ensure the cells were grown in the presence of [^{3}H]-putrescine for at least one doubling time. HT115 cells were found to have a population doubling time during exponential growth of approx 30 h and, therefore, the length of the radiolabeling period was 36 h. This time can be altered to suit the doubling time of other cell types.
4. In order to use this radiolabeling technique to monitor polyamine efflux, it is essential that, in addition to accumulating putrescine, the cells display active polyamine metabolism and convert the radiolabeled precursor, putrescine, into the higher polyamines, thereby distributing the label throughout the cellular polyamine pool. For efflux of radiolabeled polyamines to reflect normal cellular polyamine efflux, the proportions of radiolabeled polyamines need to reflect that of the total polyamine pool. HT115 cells accumulated >75% of the [^{3}H]-putrescine added. By $t = 0$ the radiolabel was distributed throughout the cellular polyamine pool in proportions that mirrored the total polyamine pool (**Table 1**). By radiolabeling cellular polyamines in this way, their release into the extracellular medium can be monitored without the need for complex purification and concentration procedures that would be essential if polyamine content of the medium was to be measured by HPLC.
5. We have overcome the problem of extracellular polyamine metabolism by using either horse serum supplemented medium, since horse serum contains little, if

any, serum amine oxidase *(26)*, or by adding 1 m*M* aminoguanidine, an inhibitor of copper amine oxidases, to the medium *(27)*. Care should be taken, however, when using the latter approach since aminoguanidine can affect both the grow of the cells and the intracellular polyamine content *(27)*.

References

1. Janne, J., Poso, H., and Raina, A (1978) Polyamines in rapid growth and cancer. *Biochim. Biophys. Acta* **473,** 241–293.
2. Pegg, A. E. (1986) Recent advances in the biochemistry of polyamines in eukaryotes. *Biochem. J.* **234,** 249–262.
3. Wallace, H. M. and Keir, H. M. (1981) Uptake and excretion of polyamines from baby hamster kidney cells (BHK-21/C13). The effect of serum on confluent cell cultures. *Biochim. Biophys. Acta* **676,** 25–30.
4. Melvin, M. A. L. and Keir, H. M. (1978) Polyamine metabolism in BHK-21/C13 cells. Loss of spermidine from cells following transfer to serum depleted medium. *Exptl. Cell Res.* **111,** 231–236.
5. Melvin, M. A. L., Wallace, H. M., and Keir, H. M. (1980) Conjugation of polyamines in mammalian cells in culture. *Physiol. Chem. Phys.* **12,** 431–439.
6. Wallace, H. M. (1987) Polyamine catabolism in mammalian cells: excretion and acetylation. *Med. Sci. Res.* **15,** 1437–1440.
7. Wallace, H. M., MacGowan, S. H., and Keir, H. M. (1985) Polyamine metabolism in mammalian cells in culture. *Biochem. Soc. Trans.* **13,** 329,330.
8. Wallace, H. M. and Keir, H. M. (1986) Factors affecting polyamine excretion from mammalian cells in culture. *FEBS Lett.* **194,** 60–63.
9. Mackarel, A. J. and Wallace, H. M. (1994) An investigation of the mechanism of polyamine efflux from human colorectal carcinoma cells. *Biochem. Soc. Trans.* **22,** 388S.
10. Wallace, H. M. and Keir, H. M. (1981) Excretion of polyamines from baby hamster kidney cells (BHK-21/C13): effect of infection with herpes simplex virus type I). *J. Gen. Virol.* **56,** 251–258.
11. Wallace, H. M., Nuttall, M. E., and Coleman, C. S. (1988) Polyamine recycling enzymes in human cancer cells. *Adv. Exptl. Biol. Med.* **250,** 331–344.
12. Wallace, H. M. and Keir, H. M. (1982) A comparison of polyamine metabolism in normal and transformed baby hamster kidney cells. *Biochem. J.* **202,** 785–790.
13. Casero, R. A. and Pegg, A. E. (1993) Spermidine/spermine N^1-acetyltransferase—the turning point in polyamine metabolism. *FASEB J.* **7,** 653–661.
14. Abdel-Monen, M. M., Ohno, K., Fortuny, I. E., and Theologides, A. (1975) Acetylspermidines in human urine. *Lancet* **ii,** 1210.
15. Russell, D. H. (1971) Increased polyamine concentrations in the urine of human cancer patients. *Nature* **233,** 144,145.
16. Rosenblum, M. G., Durie, B. G. M., Beckerman, R. C., Taussig, L. M., and Russell, D. H. (1978) Cystic fibrosis: decreased conjugation and excretion of [^{14}C]-spermidine. *Science* **200,** 1496,1497.

17. Russell, D. H., Giles, H. R., Christian, C. D., and Campbell, J. L. (1978) Polyamines in amniotic fluid, plasma and urine during normal pregnancy. *Am. J. Obstet. Gynaecol.* **132,** 649–652.
18. Russell, D. H., Durie, B. G., and Salmon, S. E. (1975) Polyamines as predictors of success and failure in cancer chemotherapy. *Lancet* **ii,** 797–799.
19. Hirsch, J. G. (1953) Spermine oxidase: an amine oxidase with specificity for spermine and spermidine. *J. Exp. Med.* **97,** 345–355.
20. Blaschko, H. and Hawes, R. (1959) Observations on spermine oxidase in mammalian plasma. *J. Physiol.* **145,** 124–131.
21. Lowry, O. H., Rosebrough, N. J., Farr, A. L., and Randall, R. J. (1951) Protein measurement with the Folin phenol reagent. *J. Biol. Chem.* **193,** 265–275.
22. Seiler, N. and Knodgen, B. (1980) High performance liquid chromatographic procedure for the simultaneous determination of the natural polyamines and their monoacetyl derivatives. *J. Chromatog.* **221,** 227–235.
23. Wallace, H. M., Nuttall, M. E., and Robinson, F. C. (1988) Acetylation of spermidine and methylglyoxal bis(guanylhydrazone) in baby hamster kidney cells (BHK-21/C13). *Biochem. J.* **253,** 223–227.
24. Pojanpelto, P. (1976) Putrescine transport is greatly increased in human fibroblasts initiated to proliferate. *J. Cell Biol.* **68,** 512–520.
25. Freshney, R. I. (1987) *Culture of Animal Cells: A Manual of Basic Techniques.* Liss, New York.
26. MacGowan, S. H., Keir, H. M., and Wallace, H. M. (1987) Presence of a tissue-type polyamine oxidase activity in mammalian serum. *Med. Sci. Res.* **15,** 687.
27. Brunton, V. G., Grant, M. H., and Wallace, H. M. (1994) Spermine toxicity in BHK-21/C13 cells in the presence of bovine serum: the effect of aminoguanidine. *Toxicol. In Vitro* **8,** 337–341.

VI

Measurement of Polyamine Effects on Cell Growth

20

Leucine Incorporation and Thymidine Incorporation

Christopher P. Denton

1. Introduction

1.1. Polyamines and Cell Growth

Although their precise role is poorly understood, polyamines appear to be important in cell growth. This is probably through their ready interaction with cell membrane components, nucleic acids, and proteins. One consequence of these interactions is that polyamines are able to modulate cell proliferation *(1)*. Moreover, regulation of polyamine levels within cells appears to be closely related to cell proliferation *(2)*, so assessment of polyamine levels and of the activity of ornithine decarboxylase activity, a key enzyme involved in polyamine synthesis, has been used in tissue extracts as an index of cell growth *(3)*. A corollary of this association between proliferation and polyamine metabolism is that basic techniques for assessing cell growth and proliferation of cells are valuable tools in the field of polyamine research *(4)*. In this chapter two methods of growth assessment that measure incorporation of radiolabeled precursors into cellular macromolecules will be described. These techniques employ the principle that cells will incorporate radiolabeled precursors of DNA and protein into newly synthesized macromolecules and the amount of incorporation will reflect both the synthetic activity of each cell in culture and also the number of cells. In many situations the amount of radionuclide incorporated can be used as an index of cell proliferation, although this is not always the case, as discussed in **Subheading 3.6.** Two of the most frequently used radiolabeled precursors are thymidine, which becomes incorporated into DNA and is generally labeled with tritium (^{3}H), and leucine, which is incorporated into most eukaryotic protein molecules. Leucine can be labeled with either

From: *Methods in Molecular Biology, Vol. 79: Polyamine Protocols*
Edited by: D. Morgan Humana Press Inc., Totowa, NJ

tritium or carbon-14 (^{14}C) and, although in this discussion [^{3}H]-labeled leucine will be considered, the same principles apply to [^{14}C]-leucine, apart from requiring different programming of the scintillation counter.

1.2. Assessment of Cell Growth

Cell growth is accompanied by the synthesis of a variety of macromolecules. Proteins are the most ubiquitous, and include the structural components of the cytoskeleton, cell membranes, and subcellular organelles, as well as essential enzymes required for metabolism and cellular signaling as well as extracellular matrix components. The rate of new protein synthesis is a reflection of the metabolic turnover of these molecules, the synthesis of new protein that accompanies an increase in size and differentiation of individual cells, and the protein requirement of cell division. The amount of new protein synthesis can be determined by measuring the incorporation of radiolabeled amino-acid substrate into cellular macromolecules during a defined labeling period ***(5)***. Radiolabeled leucine is the most frequently used substrate for such studies. This branched-chain essential amino acid is usually chosen by virtue of its ready uptake from tissue culture media by mammalian cells, the absence of *de novo* synthesis by cells that might confuse results, and its almost ubiquitous presence in cellular proteins. However, some leucine is also derived from the catabolism of other proteins, so cellular uptake is to some extent influenced by culture conditions. Radiolabeling is achieved by incorporation of either tritium (^{3}H) or carbon-14 (^{14}C) into the leucine, and these labeled amino acids are readily available from commercial sources ***(6)***. Leucine typically constitutes between 6 and 12% of amino acid residues in most mammalian protein molecules. It is however, only a minor component of some important cell proteins, which should be remembered when certain cell types are being studied. For example, it comprises only about 2% of residues in collagen, the major extracellular structural protein in most tissues, and radiolabeled proline, comprising 10–15% residues, is more appropriate if collagen synthesis, rather than overall protein production, is being investigated ***(7)***.

The quantification of new DNA synthesis is also widely used as an index of cell growth ***(8)***. DNA synthesis occurs mainly during the S-phase of the cell cycle when genomic material is being synthesized prior to cell division. The background level of new DNA synthesis is much lower and reflects the repair of damaged DNA molecules. The level of ongoing repair varies depending on cell types and culture conditions. Although the amount of reparative DNA synthesis is generally small under basal conditions, the radioactivity associated with metabolic labeling can promote DNA damage and so increase the amount of reparative DNA synthesis ***(9)***. The precise contribution of DNA turnover to thymidine incorporation can be determined experimentally using hydroxyurea to block new DNA production ***(10)***. It is important to remember that S-phase of the cell cycle does not always proceed directly to the G_1- and M-phases, so increased DNA synthesis is not necessarily accompanied by cell division and thus proliferation or growth ***(11)***. In particular, this may not occur within the time-course of an individual experiment. Therefore, thymidine incorporation data should not be directly extrapolated to changes in cell number. Nevertheless, DNA synthesis, as reflected by thymidine incorporation, is often the earliest reliable indicator of mitogenic activity for a test condition, particularly in a quiescent cell population.

Radiolabeled thymidine ([^{3}H]-TdR) is the most commonly used substrate for labeling studies of DNA synthesis. One reason for this is its specificity for DNA, so that results are not confused by alterations in RNA biosynthesis. Also, most mammalian cells are readily able to take up and utilize exogenous thymidine from culture medium via a specific salvage pathway catalyzed by the enzyme thymidine kinase. Although some discrepancies can occur in comparison with other techniques for assessing mitogenic response, such as direct cell counting or BdrU labeling ***(12)***, the relationship between thymidine incorporation and cell division is sufficiently reliable to allow its use as an index of mitogenic effects for test substances, particularly if nontransformed cells are seeded at low enough density to avoid the effects of contact inhibition at confluency, and when cells are deprived of serum and growth factors prior to addition of test mitogens.

2. Materials

1. Test cells for evaluation at confluent density in standard culture medium.
2. Serum and growth factor free culture medium, depending on cell type being studied, e.g., Dulbecco's modified Eagle's (DMEM) medium for fibroblast culture.
3. 96-Well culture plates (24, 12, or 6 well plates), 25 and 75 cm^2 plastic tissue.
4. Culture flasks and appropriate culture medium for maintenance culture of cells.
5. Microliter repeating pipet.
6. Methanol, analytical grade.
7. Phosphate-buffered saline (PBS).
8. Trichloroacetic acid, 5% (w/v).
9. Formic acid, 25*M*.
10. Tissue culture hood (conforming to BS5726-1992 for sterility of cultures and operator safety).
11. Standard tissue culture incubator (37°C/5% CO_2).
12. Water bath at 37°C.
13. Test substance under evaluation (e.g., polyamine).
14. Trypsin solution (2.5%).
15. Plastic β-scintillation vials.
16. OptiPhase safe scintillant fluid (LKB Scintillation Products, UK).

17. For leucine incorporation: tritium-labeled leucine, e.g., L-[4,5-^{3}H]-leucine, 1 mCi/mL, 51 Ci/mmol (TRAK 170B190, Amersham Life Sciences, Amersham, UK).
18. For thymidine incorporation: tritium-labeled thymidine, e.g., methyl-[^{3}H] thymidine, 1 mCi in 1 mL, specific activity 5 Ci/mmol (TRA120B374, Amersham Life Sciences).

3. Methods

3.1. Leucine or Thymidine Incorporation

Assessment of cell growth by leucine incorporation is most frequently undertaken in 96-well plates that can be seeded at either confluent or subconfluent density. Seeding density, duration of labeling, and the culture conditions will determine whether leucine incorporation is mainly reflecting protein synthesis associated with cell proliferation or simply that associated with protein turnover and metabolism in nondividing cells (*see* **Subheading 4.**). The importance of this distinction depends on the experimental aims of the study; for example, whether effects of test conditions on cell viability, cellular proliferation, or differentiation are being investigated ***(13)***. It is, however, essential that experimental conditions are standardized and that appropriate controls are included to ensure that results are interpreted correctly and to facilitate interexperiment comparisons. The greater degree of new protein synthesis occurring within cells compared with new DNA synthesis is reflected by the greater incorporation of radiolabeled leucine compared with thymidine in equivalent culture conditions (*see* **Subheading 3.6.**). The technique described in **Subheading 3.2.** allows the incorporation of tritiated thymidine or leucine into acid-insoluble cellular macromolecules (DNA or protein) to be measured in tissue-culture wells. For most experiments, 96-well plates are suitable although the protocol can be easily adapted for increased cell numbers in 6-, 12-, or 24-well culture plates.

3.2. Protocol for Tritiated Leucine or Thymidine Incorporation Assay

1. Warm trypsin and culture medium to 37°C in a waterbath.
2. Detach cells by trypsinization (time required depends on cell type and passage. Human fibroblasts typically require 5–10 min at 37°C).
3. Resuspend the cells in an appropriate volume of culture medium to give the required seeding density in the chosen culture plate i.e., in 100–200 µL for a 96-well plate (*see* **Note 1**).
4. After 24 h remove culture medium and replace with fresh serum-free medium. Generally effects on cell growth are demonstrated most clearly if the seeded cells are serum-deprived to render them quiescent, attempting to synchronize them in G_0 of the cell cycle (*see* **Note 2**).
5. Add the test substances to the individual wells, with at least five replicates of all test substance, dilutions, and controls and incubate generally overnight although time-courses should be determined for each condition.

6. Add radiolabeled leucine or thymidine (1 µL ≡ 1 µCi) to each well for a labeling period of 6–24 h (*see* **Note 3**).
7. At the end of the labeling period label-aspirate the culture medium from each well and dispose of down a designated disposal route.
8. Wash the cell monolayers three times using PBS at room temperature (200 µL/well).
9. Add 5% trichloroacetic acid (200 µL/well) and leave for 15 min to precipitate cellular macromolecules.
10. Aspirate the trichloroacetic acid from each well and discard.
11. Wash the cell monolayer with absolute methanol (200 µL/well), then leave for 5 min.
12. Add formic acid (25*M*; 200 µL/well) and leave for 5 min to solubilize the precipitated macromolecules of the cell layer.
13. Carefully transfer the contents of each well to a β vial.
14. Add 200 µL of distilled water to each well as a wash and transfer to the appropriate β vial.
15. Add 3.5 mL of scintillant fluid (e.g., OptiPhase safe) to each vial.
16. Measure the radioactivity in each sample using a liquid scintillation counter, comparing each sample disintegrations per minute with background tubes containing formic acid, water, and scintillant fluid only.

3.3. Interpretation of Results

For both leucine and thymidine incorporation data the mean dpm for replicate wells treated with the test substance, e.g., polyamine, can be compared with the mean dpm for control wells. Within experiments it is preferable to compare raw dpm figures, but the degree of interexperiment variation in radiolabel incorporated, even under similar conditions, generally makes percentage control figures more representative when independent experiments are being combined or compared. The absolute numbers of dpm are generally fewer for thymidine compared with leucine incorporation and the degree of between well and interexperiment variation greater. More replicate wells may, therefore, be required to give reliable results for thymidine incorporation experiments. Generally the total radioactivity should be at least 5–6000 dpm per well for reliable results and culture conditions; seeding density and labeling time should be adjusted if counts are consistently below this level.

3.4. Correction of Thymidine Incorporation for Viable Cell Number

Sometimes it is useful to express the thymidine incorporation per milligram of protein, performing a protein assay on a parallel experimental plate, such as the BCA protein assay ***(14)***. This is particularly useful if one or more treatment is toxic to cells so that changes in thymidine incorporation might simply reflect reduced viable cell number. The MTT assay (***[15]*** and Chapter 21) can be used in the same way, and results of these assays confirm that thymidine incorporation, particularly with quiescent cells, may not simply reflect changes in viable cell number. Thus, much of the increased thymidine incorporation probably accompanies the S-phase of the cell cycle when DNA for daughter cells is being synthesized. In view of this it is best not to equate thymidine incorporation changes with effects on cell number without other supporting evidence.

3.5. Pitfalls in Thymidine Incorporation

There are several potential pitfalls in thymidine incorporation assessments that may obscure true results. First, prolonged labeling can permit reutilization of isotope released from dying cells, giving falsely low incorporation data that underrepresent total new DNA synthesis ***(16)***. This is especially likely to be a problem in rapidly proliferating cell populations ***(13)***. Further errors may be introduced through the breakdown of the [^{3}H]-TdR and subsequent incorporation of ^{3}H into other macromolecules. Noncovalent labeling of non-DNA cellular macromolecules can also occur, leading to false overestimates of DNA synthesis ***(17)***. Also, differences between test conditions may be transient, so prolonged experiments may miss them. It has been demonstrated that thymidine itself has direct effects on cell proliferation, probably through its effects on synthesis of other DNA precursor nucleosides intracellularly ***(18)***. As with leucine incorporation data, these considerations should not differentially affect test wells within an experiment, but they should be remembered when interpreting the results obtained if standard experimental conditions are varied.

3.6. Conclusions

Both thymidine and leucine incorporation are simple and reliable tools in the assessment of cell growth. In many circumstances the two techniques give qualitatively similar results, but this is not always the case, and it may be appropriate to perform parallel experiments to compare the two techniques and allow the most appropriate to be selected for a study. Similarly, other methods for assessing viability and cell number can be investigated, such as the MTT assay discussed elsewhere in this book, and as discussed in **Subheading 3.4.** and Chapter 21, such studies may be complementary to radiolabeling methods.

Assessment of new protein synthesis by leucine incorporation is perhaps a more reliable assessment of cell growth and viability than assessing new DNA synthesis (e.g., by labeled thymidine incorporation ***[19]***), since protein synthesis reflects increased size and differentiation of individual cells and is not directly linked either to cell division or cell number ***(20)***. Nevertheless, both of these processes will affect leucine incorporation. In addition, it is essential to include all necessary control wells so experiments can be interpreted independently. Interexperiment variation often leads to differences in leucine incorporation of up to 20% for equivalent cultures and it is often more appropriate to express the incorporation in test wells as a percentage of control well dpm for interexperiment comparisons. Raw dpm counts for leucine incorporation are generally considerably greater than those for thymidine incorporation in equivalently labeled culture wells. This reflects the greater degree of protein compared with DNA synthesis under most circumstances. Leucine incorporation values are generally more reproducible between wells and experiments than thymidine incorporation data, which may reflect the fact that protein synthesis is less dependent on cell cycle stage than DNA synthesis. If test conditions are likely to alter cell viability, it is sensible to perform parallel experiments addressing this, for instance by MTT assay (*[17]* and Chapter 21), otherwise it will be difficult to distinguish whether a test condition is actually stimulating cell growth or whether it is merely not having a cytotoxic effect, so differences in leucine incorporation are simply a reflection of changes in viable cell number.

4. Notes

1. Cells can be counted using a hemocytometer slide to determine the dilution e.g., to seed 5000 cells/well in 100 µL volumes resuspend the trypsinized cells at a dilution of 5×10^4/mL. With experience it may be possible to omit the counting of cells and to simply divide the cells from a confluent monolayer in a culture flask onto a predetermined number of 96-well plates.
2. The duration of serum deprivation will depend on the cell type under investigation, but 12–48 h can be used for many cell types. Other methods of synchronization can also be employed *(11)*. If cells are not in synchronous culture the labeling time should be selected so that most cells in the culture will have undergone an S-phase during the labeling period, i.e., at least 12–16 h. It is generally best to allow the cells to adhere to the culture plastic in standard medium to ensure optimal cell viability, and then change to serum-free medium once cells are adherent and growing well. Monolayers should be washed twice (serum-free medium or sterile PBS at 37°C) before changing to serum-free medium.
3. Thymidine incorporation dpm should not be directly extrapolated to absolute amounts of DNA synthesized. If accurate determinations are required then modifications of the method can be employed using short (15 min) labeling of aliquots of a cultured cell population, determining thymidine incorporation as a percentage of total cellular protein *(21)*. Short pulse labeling studies with repeated sampling of the same cell population can also be used to evaluate synchronicity of cell cycles and to determine the duration of the S-phase in the cell cycle *(22)*. Such experiments may be valuable in designing a series of experiments employing a particular cell type to determine optimum experimental conditions.

References

1. Heby, O. (1981) Roles of polyamines in the control of cell proliferation. *Differentiation* **19,** 1–20.
2. Morrison, R. F. and Seidel, E. R. (1995) Vascular endothelial cell proliferation: regulation of cellular polyamines. *Cardiovasc. Res.* **29,** 841–847.

3. Abrahamson, M. S. and Morris, D. R. (1990) Cell type specific mechanisms of regulating expression of the ornithine decarboxylase gene after growth stimulation. *Mol. Cell. Biol.* **10,** 5525–5528.
4. Pegg, A. E. and McCann, P. P. (1982) Polyamine metabolism and function. *Am. J. Physiol.* **243** C212–C221.
5. Everhart, L. P., Hausohka, P. V., and Prescott, D. M. (1973) Measurement of growth rates and incorporation of radioactive precursors into macromolecules of cultured cells. *Methods Cell Biol.* **7,** 329–347.
6. Ball, C. R., Van den Berg, H. W., and Poynter, R. W. (1973) The measurement of radioactive precursor incorporation into small monolayer cultures. *Methods Cell Biol.* **7,** 349–360.
7. Uitto, J., Murray, L. W., Blumberg, B., and Shamban, A. (1986) Biochemistry of collagen in diseases (review). *Ann. Intern. Med.* **105,** 740–756.
8. Dover, R. (1991) Basic methods for assessing cellular proliferation, in *Assessment of Cell Proliferation in Clinical Practice* (Hall, P. A., Levison, D. A., and Wright, N. A., eds.), Springer-Verlag, London, pp. 63–81.
9. Beck, H. P. (1981) Radiotoxicity of incorporated [^{3}H]thymidine as studied by autoradiography and flow cytometry. *Cell Tissue Kinet.* **14,** 163–177.
10. Trosko, J. E. and Yager, J. D. (1974) A sensitive method to measure physical and chemical carcinogen induced unscheduled DNA synthesis in rapidly dividing eukaryotic cells. *Exp. Cell. Res.* **88,** 47–55.
11. Rew, D. A. and Wilson, G. D. (1991) Advances in cell kinetics. *Br. Med. J.* **303,** 532–537.
12. Gratzner, H. G. (1982) Monoclonal antibody to 5-bromo and 5-iodo-deoxyuridine; a new reagent for the detection of DNA replication. *Science (Wash.)* **218,** 474,475.
13. Gerecke, D. and Gross, R. (1975) Reutilisation of DNA catabolites in granulocytopoesis. *Blut* **31,** 41–48.
14. Smith, P. K., Krohn, R. I., Hermanson, G. T., Mallia, A. K., Gartner, F. H., Provenzano, M. D., Fujimoto, E. K., Goeke, N. M., Olsen, B. J., and Klenk, D. C. (1985) Measurement of protein using bicinchoninic acid. *Anal. Biochem.* **150,** 76–85.
15. Mossman, T. (1983) Rapid colorimetric assay for cellular growth and survival. Application to proliferation and cytotoxicity assays. *J. Immunol. Meth.* **65,** 55–63.
16. Maurer, H. R. (1981) Potential pitfalls of ^{3}H-thymidine techniques to measure cell proliferation. *Cell Tissue Kinetics* **14,** 111–120.
17. Schneider, W. C. and Greco, A. E. (1971) Incorporation of pyrimidine deoxyribonucleotides into liver lipids and other components. *Biochim. Biophys. Acta* **228,** 610–626.
18. Ooi, S. O., Sim. K. Y., Chung, M. C., and Kon, O. L. (1993) Selective antiproliferative effects of thymidine. *Experientia* **49,** 576–581.
19. Boulton, R. A. and Hodgson, H. F. (1995) Assessing cell proliferation: a methodological review. *Clin. Sci.* **88,** 119–130.

20. Simon, J. S., Baum, J. S., Moore, S. A., and Kasson, B. G. (1995) Arginine vasopressin stimulates protein synthesis but not proliferation of cultured vascular endothelial cells. *J. Cardiovasc. Pharmacol.* **25,** 368–375.
21. Gospodarowicz, D. and Moran, J. S. (1975) Determination of ^{3}H incorporation into DNA in vitro. *Methods Cell Biol.* **3,** 320–329.
22. Stubblefield, E. (1968) Synchronisation methods for mammalian cell cultures. *Methods Cell Physiol.* **3,** 25–44.

21

Tetrazolium (MTT) Assay for Cellular Viability and Activity

David M. L. Morgan

1. Introduction

In the course of examining the effects on cells of polyamines, their metabolites, and polyamine analogs, it is often necessary to make some measure of cellular activity as an indicator of cell damage or cytotoxicity. One of the simplest assays utilizes 3-[4,5-dimethylthiazol-2-yl]-2,5-diphenyl tetrazolium bromide (MTT), a water-soluble yellow dye that is readily taken up by viable cells and reduced by the action of mitochondrial dehydrogenases *(1)*. The reduction product is a water-insoluble blue formazan, that must then be dissolved for colorimetric measurement. Ethanol or propanol *(2)*, acid-isopropanol (0.04*M* HCl in propan-2-ol) *(3)*, acid-isopropanol plus 10% Triton X-100 *(4)*, mineral oil (unspecified), or dimethyl sulfoxide (DMSO) *(5)* have all been suggested. In the author's hands DMSO was found to be the most satisfactory (**Fig. 1**).

Formazan production is directly proportional to cell number over the range ~200–50,000 cells/well *(3)*, and can be used to detect the presence of very small numbers of living cells. Metabolically inactive cells, such as erythrocytes, do not produce significant amounts of formazan. Conversely, the amount of formazan produced per cell in a given time is dependent on the metabolic activity of that cell; activated lymphocytes generate up to ten times as much formazan as resting cells *(3)*. Formazan production is linear with time over 30 min–2 h *(2,3,6)*, and is also proportional to MTT concentration, the relationship differing with different cell types *(6)*. Measurements of formazan production show good correlation with [^{3}H]thymidine uptake assays *(2,3)* and, despite a report to the contrary *(4)*, in the author's hands the method works equally well with adherent and nonadherent cells. The MTT procedure assesses the activity (and number) of living cells at the end of an experiment, whereas [^{3}H]thymidine

From: *Methods in Molecular Biology, Vol. 79: Polyamine Protocols*
Edited by: D. Morgan Humana Press Inc., Totowa, NJ

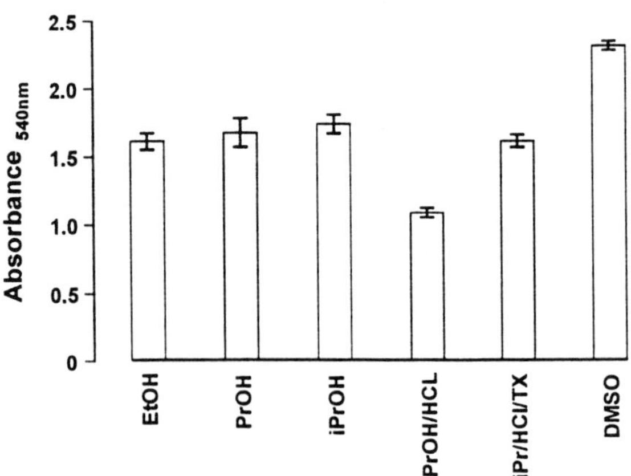

Fig. 1. Effect of solvent on formazan absorbance. EtOH, ethanol; PrOH, propanol; iPrOH, 2-propanol; iPrOH/HCl, 0.04*M* HCl in 2-propanol; iPrOH/HCl/TX, 0.04*M* HCl in 2-propanol plus 10% Triton X-100; DMSO, dimethyl sulfoxide (error bars show S.D., $n = 5$).

incorporation measures DNA synthesis over a much longer period (*see* Chapter 20). A further advantage of the MTT assay is that tetrazolium-cleaving enzymes are not present in serum. A disadvantage is that this is a destructive technique in that the cells cannot subsequently be used for any other assay. However, the culture medium can be saved for other assays if required.

2. Materials

1. Flat-bottomed 96-well tissue culture grade microwell plates with lids (e.g., 1-67008A, Life Technologies Ltd., [Paisley, UK]; 655180, Greiner [Frickenhausen, Germany]; 76-003-05, ICN Pharmaceuticals Inc, Thame, UK, and Costa Mesa, CA).
2. Repetitive pipet, e.g., Multipet (Eppendorf 4780, Merck Ltd., Lutterworth, UK).
3. Combitip, 1 and 5 mL, for Multipet (e.g., 307/8534/02 and 307/8534/03, Merck Ltd; NP7321 and NP7331, Alpha Laboratories [Eastleigh, UK]).
4. Multichannel pipets ***(2)***, e.g., Proline 8-channel (Alpha Laboratories), if required (*see* **Subheading 3.**).
5. Plate reader equipped with a 540-nm filter and linked to computer or printer.
6. For cells grown in suspension, a centrifuge capable of accepting 96-well culture plates is required.
7. 3-[4,5-Dimethylthiazol-2-yl]-2,5-diphenyl tetrazolium bromide (MTT; Sigma Chemical Co. Ltd, Poole, UK); store at –20°C.
8. 1-[4,5-Dimethylthiazol-2-yl]-3,5-diphenylformazan (MTT formazan; if required, *see* **Subheading 4.**).
9. Dimethylsulfoxide (DMSO).
10. Culture medium appropriate to cell type used.

11. MTT stock solution, 5 mg/mL. Dissolve MTT in serum-free culture medium and store in suitable size aliquots (400 μL is sufficient for one plate) at –20°C.
12. MTT formazan standard, 1 mg/mL in DMSO (if required, *see* **Subheading 4.**).
13. MTT reagent. Immediately before use thaw out sufficient MTT stock solution and dilute 1 to 10 with complete (i.e., serum-containing) culture medium (final concentration of MTT is 0.5 mg/mL; *see* **Note 1**).

3. Methods

3.1. Adherent Cells

1. Dispense 100–200 μL of an appropriate cell suspension, in the appropriate culture medium, into the inner 60 wells of a 96-well flat-bottomed tissue culture plate (*see* **Note 2**), the outer 36 wells of which each contain 200 μL of PBS (for some reason cell growth is less reproducible in the outer wells).
2. Allow the cells to adhere and grow, to confluence if required, under standard culture conditions.
3. Apply treatments and so forth, for the required times.
4. Rapidly invert the plate over a sink or other suitable container and shake (or if it is necessary to retain the medium, remove it using a multichannel pipet).
5. Keep plate inverted and blot on paper towel.
6. Add 50 μL of MTT reagent, using a Multipet equipped with a 1-mL combitip, to all sample wells and to wells A1–H1 (blanks), omitting wells to be used for protein estimation (if required, *see* **Subheading 4.**).
7. Incubate the plates at 37°C for appropriate time (*see* **Note 3**).
8. If required, add 5, 10, 15, 20, and 25 μL of the MTT formazan standard (containing 5, 10, 15, 20, and 25 μg of formazan) to triplicate outer wells of the 96-well plate, starting at well A2 (*see* **Note 4**), plus sufficient DMSO to make the volume in each well up to 50 μL.
9. Add 100 μL DMSO to each well (samples and blanks, but omitting wells to be used for protein estimation), using a Multipet with a 5 mL Combitip.
10. Shake plates gently for 15 min on an orbital shaker (*see* **Note 5**).
11. Measure the absorbance of each well at 540 nm on plate reader (*see* **Note 6**). If the plate reader is connected to a computer then the data can be saved as a text file and later imported to a spreadsheet if required.

3.2. Suspension Cells

1. Centrifuge plate at low speed (~800*g* for 10–15 min *[2,4]*).
2. Aspirate medium, taking care not to lose any cells.
3. Treat as for adherent cells.

3.3. Calculation of Results

Controls and samples should be on the same plate. If more than one plate is used, then controls should be replicated on all plates. Formazan generation in the test wells can then be expressed as percentages of that in control wells. If

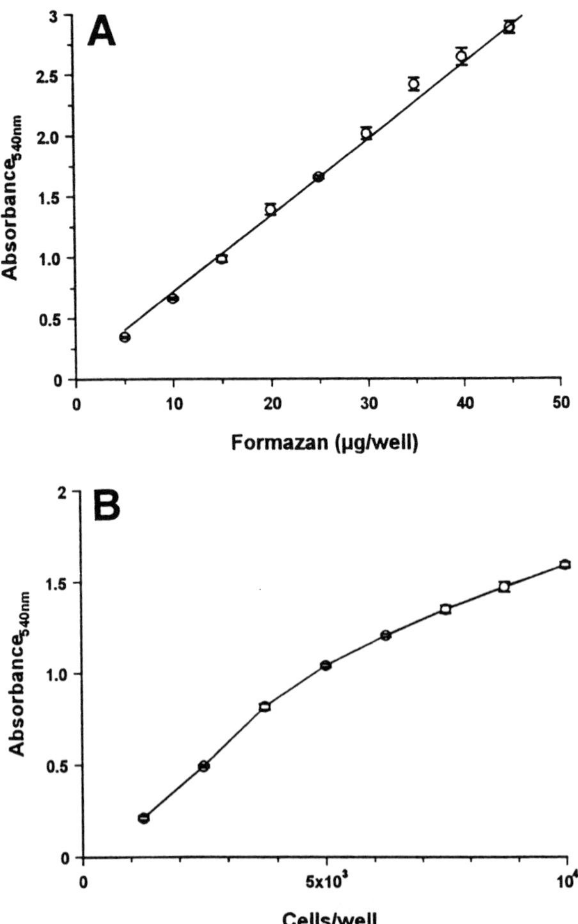

Fig. 2. Linearity of MTT assay. **(A)** Formazan concentration (error bars show S.D., $n = 3$). **(B)** Cell number (ECV304 cells, plated 24 h; error bars show S.D., $n = 6$).

formazan standards have been included, and cell protein is determined in parallel treated and control wells, formazan production can be quantitated in terms of µg formazan produced/h/µg protein.

4. Notes

1. MTT should be added in complete medium because formazan generation is dependent on extracellular D-glucose *(6)*; omission of serum is unnecessary because the small volume of medium used (40 µL) does not cause turbidity on addition of DMSO.
2. Because of their optical properties, flat-bottomed culture plates should be used for both adherent and nonadherent cells.

3. Time of incubation with MTT must be determined individually for different cell types and experimental conditions and is usually 1–3 h. As a guide, for J774 cells (a murine macrophage-like cell line) plated at ~500,000 cells/mL, a 1-h incubation will result in absorbancies of 1.1–1.4 at 540 nm; ECV304 cells (a human umbilical vein endothelial cell line), at ~50,000 cells/mL will have an absorbance of 1.3–1.5 after 1 h.
4. The absorbance of MTT formazan in DMSO is linear with concentration from 5–50 μg/well, but at 50 μg/well is >2.5 (**Fig. 2A**).
5. The orbital shaker should be carefully adjusted because too vigorous a movement will result in splashover from one well to the next: If the shaking is too gentle there may be insufficient mixing.
6. Some authors ***(2–4)*** have suggested that absorbance measurements should be made at two wavelengths, 540–570 nm and 620–690 nm, and the difference used in calculations. In the author's view, after comparative tests, this is unnecessary if ≤50 μL of medium is present on addition of DMSO.

References

1. Slater, T. F., Sawyer, B., and Strauli, U. (1963) Studies on succinate-tetrazolium reductase systems. III. Points of coupling of four different tetrazolium salts. *Biochim. Biophys. Acta* **77,** 383–393.
2. Denizot, F. and Lang, R. (1986) Rapid colorimetric assay for cell growth and survival: modifications to the tetrazolium dye procedure giving improved sensitivity and reliability. *J. Immunol. Methods* **89,** 271–277.
3. Mossman, T. (1983) Rapid colorimetric assay for cellular growth and survival: application to proliferation and cytotoxicity assays. *J. Immunol. Methods* **65,** 55–63.
4. Takeuchi, H., Baba, M., and Shigeta, S. (1991) An application of tetrazolium (MTT) colorimetric assay for the screening of anti-herpes simplex virus compounds. *J. Virol. Methods* **33,** 61–71.
5. Carmichael, J., DeGraff, W. G., Gazdar, A. F., Minna, J. D., and Mitchell, J. B. (1987) Evaluation of a tetrazolium-based semiautomated colotimetric assay: assessment of chemosensitivity testing. *Cancer Res.* **47,** 936–942.
6. Vistica, D. T., Skehan, P., Scudiero, D., Monks, A., Pittman, A., and Boyd, M. R. (1991) Tetrazolium-based assays for cellular viability: a critical examination of selected parameters affecting formazan production. *Cancer Res.* **51,** 2515–2520.

Index